FIXED BED OPERATIONS

Flow Distribution and Efficiency

ELLIS HORWOOD SERIES IN CHEMICAL ENGINEERING

Series Editor: Dr. DAVID SHARP, OBE, lately General Secretary, Society of Chemical Industry, formerly General Secretary, Institution of Chemical Engineers, and former Technical Director, Confederation of British Industry

THERMODYNAMICS FOR CHEMIST AND CHEMICAL ENGINEERS
M. H. CARDEW
ENVIRONMENT AL BIOTECHNOLOGY
C. F. FORSTER and D. A. J. WASE
HEAT EXCHANGE ENGINEERING Volumes 1 & 2
E. A. FOUMENY and P. J. HEGGS
POWDER TECHNOLOGY AND MULTIPHASE SYSTEMS:
Gas Permeametry and Surface Area Measurement
G. J. I. IGWE
CHEMICAL REACTORS FOR GAS—LUQUID SYSTEMS
F. KAŠTÁNEK, J. ZAHRADNÍK, J. KRATOCHVÍL and J. ČERMÁK
FLUIDIZATION: Idealized Systems with Applications
M. KWAUK
MAJOR CHEMICAL HAZARDS
V. MARSHALL
BIOLOGY AND BIOCHEMISTRY FOR CHEMISTS AND CHEMICAL ENGINEERS
W. J. MITCHELL and J. C. SLAUGHTER
INSTRUMENTATION AND AUTOMATION IN PROCESS CONTROL
M. J. PITT and P. E. PREECE
FIXED BED OPERATIONS: Flow Distribution and Efficiency
V. STANĚK
MASS TRANSFER
J. A. WESSELINGH and R. KRISHNA
BIOLOGICAL TREATMENT OF WASTE-WATER
M. WINKLER

Ellis Horwood Limited also publishes related titles in
APPLIED SCIENCE AND INDUSTRIAL TECHNOLOGY
CHEMICAL AND INDUSTRIAL CHEMISTRY
ENGINEERING SCIENCE
BIOLOGICAL SCIENCES

FIXED BED OPERATIONS

Flow Distribution and Efficiency

V. STANĚK Ph.D

Institute of Chemical Process Fundamentals,
Academy of Sciences of the Czech Republic, Prague

Translation Editor
DAVID SHARP, OBE
lately General Secretary, Society of Chemical Industry,
formerly General Secreatary, Institution of Chemical Engineers,
and former Technical Director, Confederation of British Industry

ELLIS HORWOOD
NEW YORK LONDON TORONTO SYDNEY TOKYO SINGAPORE

English edition first published in 1994 in coedition between
ELLIS HORWOOD LIMITED
Market Cross House, Cooper Street,
Chichester, West Sussex, PO19 1EB, England
A division of
Simon & Schuster International Group
and
ACADEMIA
Publishing House of the Academy of Sciences
of the Czech Republic, Prague, Czech Republic

Distributed in the East European Countries,
China, Cuba, Mongolia, North Korea and Vietnam by
Academia, Publishing House of the Academy of Sciences
of the Czech Republic, Vodičkova 40, 112 29 Praha 1,
Czech Republic

Printed in the Czech Republic

British Library Cataloguing in Publication Data

Staněk, V.
Fixed bed operations: flow distribution and efficiency. —
(Ellis Horwood series in chemical engineering)
I. Title II. Series
660
ISBN 0–13–322033–8

Library of Congress Cataloging-in-Publication Data

Staněk, Vladimír.
[Vliv nerovnoměrného toku tekutin na procesy ve vrstvách pevných částic. English]
Fixed bed operations: flow distribution and efficiency / V. Staněk;
translation editor, David Sharp.
p. cm. — (Ellis Horwood series in chemical engineering)
Includes bibliographical references and index.
ISBN 0–13–322033–8
1. Multiphase flow. 1. Title. II. Series.
TA357.5.M84S713 1992
660 284—dc20 91—45985
CIP

CONTENTS

1. Introduction

Multiphase systems in which a layer of solid particles is permeated by gas, liquid or both, are among the most widely used in a variety of industrial activities. As examples one can cite packed distillation and rectification columns, absorption and scrubbing equipment, trickle bed catalytic reactors, biological filters for waste water treatment, reactors with immobilized enzymes, oil shale retorts, adsorption columns for separation of noxious contaminants, preparative gas and liquid chromatography columns, nuclear pebble bed reactors, metallurgical equipment such as blast furnaces and agglomeration machines, hydrogen accumulators on the basis of intermetallides, reactors for oxidation of organic contaminants in waste gases and exhaust gases of motor vehicles (honeycomb reactors), heat exchangers, water cooling towers, and others.

Concurrently with the industrial application of fixed bed systems, which were introduced in the last century, systematic research into their properties has been carried out. The earliest separation columns were packed with coke and were inexpensive and their weight per unit volume was relatively low. Contemporary modern structured packings are developed with the aim to optimize the structure of the flow of phases and to minimize the pressure drop per mass transfer unit. Research into real flows through the layer of particles extended into the field of reactor engineering, in particular, fixed bed and trickle bed catalytic reactors.

The structure of the bed of particles as well as the velocity and pressure fields deviate in real systems from the idealized, mostly homogeneous structures. The theory of processes and design procedures for fixed bed systems with ideal homogeneous layer and ideal homogeneous (uniform) flow has been worked out nearly to perfection. Also the applications of underlying technological operations in industrial equipment under these idealized conditions have been well known. It can therefore be said that the potential for further progress of technological equipment with a fixed layer of particles and homogeneous flow has been virtually exhausted.

Thanks to the advanced computers it is now possible in the description of the phenomena of the processes involving layers of particles to discard almost all simplifications hitherto dictated by foreseen mathematical difficulties. Distribution properties of the system and nonlinearities usually rule out analytical solutions and approximate linearized solutions usually do not reflect the more complex properties and interactions within the system.

The possibility of describing the system on the basis of clearly defined physical properties with distinct departure from pure empiricism has been significantly enhanced. This last property is particularly important for a successful application of the theory of similarity, for scale up, as well as for the choice of suitable independent experiments devised for parameter evaluation.

The aim of this book is to show the reader the relations between structural inhomogeneities of the random layer of particles, the inhomogeneities appearing in the velocity and pressure fields and, in turn, the changed energy requirements for the flow, intensity of heat transfer, intensity of mass transfer and the course of the chemical reaction.

A significant part of this endeavour is the unified view of the role of inhomogeneities in affecting transport phenomena and chemical kinetics in layers of particles, distinct from the technological equipment. In practice we seek to work out a general theoretical basis which takes into account the fact that the particles are, for example, catalyst pellets, iron ore particles or the packing of a heat regenerator, and, similarly, that the flowing fluid may be hydrogen, liquid iron, or absorption liquid. We endeavour to establish appropriate physico-chemical properties of the fluid, dimensions and shape of the particles and appropriate constraints following from the temperature and pressure dependences of physical and chemical properties.

1.1. NATURE OF PROCESSES IN LAYERS OF PARTICLES

The wide application of equipment with a fixed bed of particles has been stimulated by some of its characteristic features. In particular it is the large interfacial fluid/solid or gas/liquid surface which in large separation columns contributes to the reduction of the height of a transfer unit and in catalytic reactors to suppresion or elimination of the effects of external diffusion. Individual elements of the fluid exhibit on passing through the layer a relatively narrow distribution of the residence times and the flow thus approaches plug flow. This property is of particular importance for the efficiency of the separation equipment and selectivity of chemical reactions. A low degree of mixing of the fluid in the direction transverse to the principal direction of the flow is often a drawback, particularly if a heat transfer into or out of the layer is needed. This property contributes also to the fact that radial temperature and concentration profiles appearing due to the flow inhomogeneities persist in a large part of the equipment.

Inhomogeneities of the structure of the bed, of the flow of phases, heat transfer and chemical reactions, however, do not act in isolation. On the contrary, complex interactions appear among them. Possible interactions in systems with solid particles and flowing phase(s) are schematically shown in Fig. 1-1. The STRUCTURE OF THE BED of particles, written on the top of this scheme is given by the spatial distribution of the void fraction and/or particle size. The inhomogeneities of the structure will directly cause inhomogeneities (nonuniformities) of the VELOCITY AND PRESSURE fields (due to the nonuniform spatial distribution of resistance to the flow) and HEAT DISPERSION and CHEMICAL REACTIONS (at least due to the changed solid/fluid surface area). The CHEMICAL REACTIONS represent here gas/solid, fixed bed and trickle bed catalytic reactions. These effects are shown schematically by the three leftmost arrows.

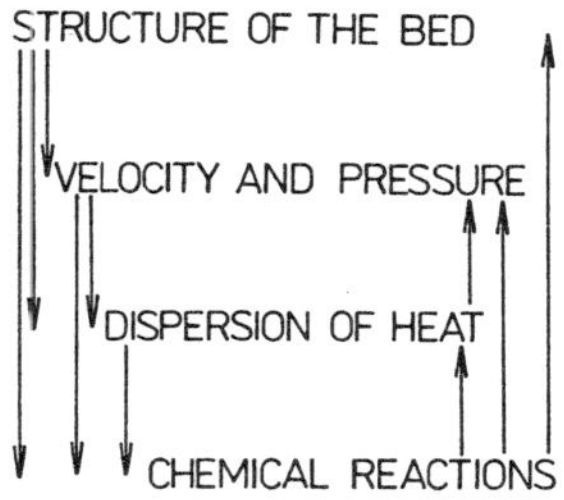

Fig. 1-1
Scheme of interactions in a system with solid particles

The nonuniformities of the VELOCITY AND PRESSURE field will in turn induce additional changes in HEAT DISPERSION and CHEMICAL REACTIONS through the convective contributions; see the two downward pointing arrows originating from the VELOCITY AND PRESSURE symbol.

The changes in the DISPERSION OF HEAT induced by the two above effects will further cause additional changes in the course of CHEMICAL REACTIONS due to the temperature dependent rates of chemical reactions as shown by downward pointing arrows originating from the DISPERSION OF HEAT symbol.

The three upward pointing rightmost arrows originating from the CHEMICAL REACTIONS symbol indicate the reverse effect of chemical reactions on the DISPERSION OF HEAT due to the changed rate of evolution (absorption) of the reaction heat, on the VELOCITY AND PRESSURE FIELD through the changed physico-chemical properties of the fluid and on the STRUCTURE OF THE BED through eventual thermomechanical disintegration of particles or formation of deposits. Even in the absence of chemical reaction there is an effect of HEAT DISPERSION on the VELOCITY AND PRESSURE field through the temperature changes of physical properties of the fluid, see the fourth upward pointing arrow.

Comprehension of the real structures and real flows through such layers, including the consequences for transport phenomena and chemical reactions, however, is not merely a question of better quantitative agreement or higher precision of the model. Structural inhomogeneities give rise also to qualitatively new phenomena important both for the progress of theoretical knowledge and industrial practice.

Undesirable phenomena appearing as a consequence of inhomogeneities have to be foreseen, correctly interpreted and reflected, or, eventually suppressed by appropriate measures, in the design of the equipment or in the control algorithm. In other cases the purposeful structuralization of the bed can provide for energy savings or a more advantageous course of the underlying technological process.

2. Structural inhomogeneities of random fixed beds

2.1. CLASSIFICATION AND CHARACTERISTICS OF STRUCTURAL INHOMOGENEITIES OF THE BED

An important factor in a number of processes taking place in a layer of particles is the formation of a large interfacial surface either directly between gas and solids, or, indirectly between gas and liquid while the orientation of the surface of particles is more or less secondary. In these cases the particles themselves are usually dumped into the apparatus and we then speak of random beds.

Random beds may be characterized by the particle size (monosized layers), or, eventually, by the distribution function of particle size, and by the void fraction.

The volume void fraction is defined as the volume of the voids expressed as a fraction of the total volume of the voids and the particles. Similarly, one can relate [45] the void fraction to the conditions at a given cross section and define the area void fraction as a fraction of the free cross section to the total cross sectional area.

In abstraction when one views the bed as a continuum (the pseudohomogeneous model approach) one can work, in relation to the reality, with several notions:

Bulk mean voidage - Fractional free volume in the whole bed, or sometimes in an infinite bed free of the effect of containing surfaces.

Local voidage - Fractional free volume in an element of the bed volume in the form of a thin slab, cylindrical or spherical shell. As the width of the element tends to zero, local voidage tends towards local area voidage.

Point voidage - In real beds that are not really continua, the point voidage is a two-value function of the coordinates assuming values zero or unity.

Local mean voidage - The fractional volume in a local region with reference to a point in space.

Local voidage of a spherical shell, as a function of the radial distance from the center, has been found to exhibit [141] damped oscillations, similar to a bed of spheres in the proximity of the wall. These oscillations, however, decay faster.

The frequency function of the distribution of local mean void fraction is interesting in that it has been found [142] to be similar to the frequency distribution function of other local mean quantities such as hydraulic radius, stagnant mixing and diffusion film thickness around particles.

From the above it is seen that in the study of beds of particles on the basis of pseudohomogeneous models, assuming the properties of the continuum, one has to be very careful when applying the concept of a "differential" element of the bed. This is particularly important when comparing theoretical results computed as solutions of the differential equations derived on the basis of pseudohomogeneous models with "local" measurements of corresponding quantities.

It is apparent that in this respect the "differential" volume has to be small compared to the bed scale (e.g. bed diameter) and large compared to the particle scale (e.g. particle diameter).

A priori it may be evident that although particles in a random bed are located randomly, this randomness needs a suitably precise definition in order that we may quantify eventual deviations from the randomness. The random layer may be defined [45] as a layer in which particles of equal size and shape have equal probability to occupy an arbitrary position within the bed.

Consequently, as follows from this definition, the area void fraction in an arbitrary cross section of the random bed is the same and, in addition, equal to the bulk void fraction. Local void fraction is then, apart from the limitations stemming from the dimensions of the volume element, independent of the position within the bed.

For a random bed formed by spheres of equal diameter, d, one can derive, on the basis of the study of Wicksell [44], the following frequency function, $z(\varphi)$, of the distribution of diameters, φ, of discs appearing in the cross section through such a random layer

$$z(\varphi) = \frac{\varphi}{d(d^2 - \varphi^2)^{1/2}} \tag{2-1}$$

From this function one can easily find that the mean area of a single disc appearing as a result of sectioning the spheres in the bed by an arbitrary plane equals $(\pi\, d^2/6)$, i.e. 2/3 of the maximum sphere cross section.

A perfectly random layer of equal sized particles thus has in each volume element the same void fraction, the same solid/fluid interfacial surface and particles of the same size. These properties are important from the standpoint of chemical and physical processes taking place in the layer passed by one or more fluid phases. Tondeur and Kvaalen [145] showed that in a given contacting or separation device with a given transfer surface the best configuration is that which gives the most uniform distribution of the entropy production. In practical situations, however, the structure of the bed affects the distribution in time and space of both the driving forces and the surface area. This aspect will be subject of subsequent paragraphs and chapters.

Theoretical studies of random beds concentrate mostly on beds of spherical particles where geometrical considerations provide a powerful tool. There are, of course, other geometries [140], as e.g. cubes etc., that could be also used but their practical significance is limited. For a review see e.g. Ziolkowska and Ziolkowski [143].

In many cases the studies start from regular packing configurations made up of layers, rows or rhombs of spheres. In fact, there is a finite and not very large number of alternatives in which the monosized spheres can be stacked.

Let us note that a regular rhomboedral structure of monosized spheres, where the layers spacing is only $d\sqrt{2/3}$ exhibits the lowest void fraction 0.2595. The orthorhombic packing, where the spacing of individual layers of spheres is d has the void fraction 0.3954. These values may be later useful for comparison with the data for random beds of spheres which are of maximum practical interest.

Experimental, theoretical and computer investigations of the void fraction in random beds have revealed [46,47,134,135,141] that there exist two extreme states of a randomly packed bed termed the random dense and the random loose packing. Random dense beds result when the bed, after filling by particles, is exposed to shaking or certain type of vibrations reducing its void fraction.

Random loose packings result when shaking is not applied, from generally "gentle" methods of filling the bed, or, from loosening the dense bed.

These two states of the random bed are characterized by fairly well reproducible values of the void fraction. These values, of course, depend on the aspect ratio, i.e. the ratio of the diameter of the sphere to the diameter of the vessel. Experimental data newly extrapolated to infinite radius of the vessel and infinite depth of the packed layer set the values of the void fraction at 0.36(3) and 0.39(9) for the dense and the loose packing respectively. Gotoh and Finney [139] speculated on the basis of a statistical geometrical approach that the value of the void fraction for the loose packing is the upper stability limit for the void fraction of a random bed of spheres. The void fraction of a bed of monosized spheres at incipient fluidization is about 0.46.

The reproducibility of a given method of packing is fairly high and evaluation of a large number of experiments shows normal distribution of the resulting void fractions about the mean with the standard deviation of only about 0.0016 void fraction units. Let us note also that the mean value for the loose random packing is quite close to the regular orthorhombic packing in spite of the obvious dissimilarities.

An important characteristic of a packing is the coordination number. For spheres it is defined as the number of spheres in contact with any given sphere. Numerical values of the coordination number can be also evaluated on the basis of geometrical models for regular beds of spheres, and, in particular, for the rhomboedral and the orthorhombic structures where they amount to 12 and

8 respectively. Experimentally, the coordination number is relatively hard to determine as it is often difficult to distinguish between a contact and a near--contact point.

The number of contact points directly affects mechanical properties of the bed (cohesion, force transmission) and, indirectly, for instance, the effective heat conductivity of the bed at zero velocity of the fluid, or the static hold-up of liquid in a trickle bed.

Generally, one can say that the coordination number is correlated with the void fraction and increasing void fraction of the bed, i.e. going from the dense to loose random beds, will cause a decrease in the coordination number. The theoretical relationship between the coordination number, C_N, and the void fraction of a bed of uniform size spheres has been given by Ouchiyama and Tanaka [131]:

$$C_N = \frac{32}{13}(7 - 8\varepsilon) \tag{2-2}$$

Regular beds, of course, have the same coordination number for each sphere (individual coordination number), which is not the case in random beds. Nevertheless, it has been found that on changing over to more compact packings the distribution (spread) of the coordination number becomes narrower. Rather surprisingly the effect of container walls on the coordination number is not very significant. In other words, the number of contact points per sphere does not vary appreciably with the radial distance from the container wall.

Also the distribution of the local mean voidage [141] grows broader for lower coordination numbers, and, as mentioned above, shifts towards higher local mean voidages. The distribution curve is distinctly skewed, particularly for narrow type distributions, due to the existence of a definite lower limit for the voidage while the upper limit varies with the coordination number.

The factors that affect the void fraction in a random bed can be divided essentially into two categories: Factors associated with the particles themselves and with the method of packing.

Factors affecting the void fraction associated with the particles themselves.

Among the most important factors is the size (diameter) of the particles. Truly random beds can exist only in infinite dimensions, while in practice the bed is always confined in a vessel. In this respect the effect of particle size can be taken as a relative one, i.e. in the form of the aspect ratio of the particle to the bed diameter.

However, the absolute particle size also plays a role as it has been observed that with diminishing particle size the surface forces become increasingly important. This is of importance for chromatographic columns where phenomena such as particle clustering appear. In general one can say that for the same aspect ratio the beds with smaller particles will tend to exhibit higher void fractions.

For beds of monosized spheres the effect of the aspect ratio is usually dominant over the effect of the method of packing. Zagrafskaia et al. [78] recommended a simple correlation for the dependence of the void fraction on the aspect ratio

$$1 - \varepsilon = (1 - \varepsilon_{\infty}) \left[1 - \left(\frac{r}{R} \right)^2 \right] \tag{2-3}$$

The effect of particle shape must also be taken into account. Highly nonspherical particles such as e.g. Raschig rings, Berl saddles, full and hollow cylinders, catalyst pellets, differ substantially in the mean void fraction of the resulting bed. Theoretical analysis is difficult because the orientation of the particle in space now also plays a role. From this stems also a much stronger effect of the packing method on the resulting void fraction. In some cases there are not only contact points among particles and with the container walls but also contact lines and catalyst surfaces.

"Technical" particles such as ore or coke often do not deviate in sphericity as much as the particles previously mentioned do and some of the theories derived for spheres may be applied. In both cases a way to incorporate the effect of the shape is to introduce the shape factor into the expressions based on the assumptions of spherical particles. An appropriate form is usually a correction on the apparent surface of the particle. The shape factor is then used in conjunction with the nominal particle size or equivalent spherical particle diameter in expressions derived for spheres. This ensures corresponding surface area to particle volume ratios. The shape factor proved to be an effective tool in dealing with both non-spherical particles as well as polydispersed particles, see e.g. Molerus and Schweinzer [226].

Roughness of the surface of particles and the walls of the container also affect the void fraction. This effect is particularly important for technical particles such coke, ore, and agglomerates which tend to bridging and arching. The effect of the roughness is then so severe that one cannot distinguish sufficiently precisely between the effect of imperfect shape and the effect of surface roughness.

The shape of the containing vessel also plays a role in affecting the void fraction of the bed. The effect of cylindrical walls and plane containing surfaces will be dealt with in greater detail subsequently. The effect of corners if the container is of rectangular cross section requires special attention. The void fraction there will be markedly increased.

Factors affecting the void fraction associated with the method of packing

The factors affecting the void fraction associated with the method of packing are of particular importance for non-spherical particles. Irregularly shaped particles form beds in a different manner from spheres and the result is considerably more affected by the method of packing. Packing methods generally differ by intensity of deposition of particles, the speed of filling the bed and the kinetic energy of a particle deposited in the bed. The trend of the effect of these factors depends also on the mechanical properties of particles and the way they convert the kinetic energy. Another factor that will affect the result is whether the vessel to be packed is filled by liquid or empty.

The void fraction of the bed immediately after filling can be further modified by, for instance, pressing. These procedures are usually applied in packing chromatographic columns with small particles. To date there exist in principle two methods of bed compression: Axial, using a plunger, and radial, applying pressure externally on the deformable column wall (plastic cartridge).

For larger particles one can apply, for instance, vibration. Vibrated beds tend to be more closely packed and the effect is usually more pronounced for irregular particles, where vibration helps re-orientation of the particles. The effect of vibration is usually more effective with higher frequencies and smaller amplitudes. Higher amplitudes tend to loosen the bed.

In real industrial equipment with a fixed bed of particles, of course, deviations from perfect randomness practically always occur. From the standpoint of the cause for their occurence one can classify these deviations into the following categories:

a) Deviations due to the presence of surfaces confining the bed spatially.

b) Deviations due to segregation, agregation, clustering and arching during preparation of the bed.

c) Deviations due to the different size of particles, their attrition, erosion and fouling of the bed.

d) Deviations in regions of contact of particles of different size and/or shape.

a) In technological equipment the particles are placed in various containers, columns, reactors, furnaces, etc and the bed is thus spatially contained both in the horizontal direction by the walls as well as in the vertical direction from below by the supporting grid or bottom.

The radius of curvature of the confining surfaces is substantially greater than the radius of curvature of the particles themselves, so that these differences bring about changes in the frequency function, Eq.(2-1), of the distribution of the radii of discs in the plane near the confining surface.

Because the top surface of the bed is also usually levelled and in some cases compressed, analogous changes in the distribution function also appear near the top of the bed so that this surface has to be, in principle, also considered as a wall although physically it is not a confining surface.

Into the same category fall also the deviations from randomness in the neighbourhood of thermocouple wells, cooling coils, fins, etc.

b) Deviations due to segregation during charging are of two kinds: If the layer is formed by a mixture of particles which are not of the same size the segregation takes place during the column filling with the tendency of the larger particles to occupy the regions near the wall. This phenomenon has been observed even in chromatographic columns with very small particles and very narrow distribution of their size. In blast furnace charges with the presence of fines the segregation is also considerable with the fines being present mainly in the central zone thereby contributing to further decrease of the void fraction there. In blast furnaces it has also been observed that many materials breakdown even at low temperatures and tests have been developed [165] to assess this property.

If, on the contrary, the bed is to consist of different particles charged into the bed separately so as to create zones in the bed of particles of different size, desegregation of particles during chargings occurs giving rise to zones with mixtures of the particles. This is the case of the blast furnace where the principal components charged, i.e. coke and ore, are of substantially different dimensions and are charged separately. In this case, of course, desegregation and particle mixing also takes place during descent of the particles through the shaft of the blast furnaces.

Agregation and clustering is also of importance for small particle beds (chromatographic columns). These phenomena give rise to two types of paths through the bed. One type is a path on the scale of particle size, the other between clusters of particles. Although the fractional volume of the pores on the scale of clusters need not be large, their flow resistance is substantially different with highly detrimental effect on the efficiency of the column.

Arching, on the other hand, is characteristic of large particle beds and of particles with coarse surface. Its result are voids disproportinately large compared to particle size.

c) Inhomogeneities of the structure of presumably monodisperse particle beds are given rise to, in some cases, even during filling by particle attrition and the production of fines. Fines, in addition, occur during operation due to particle erosion by the flowing phases and, further, as a consequence of thermomechanical strain. Temperature differences induced secondarily during, for instance, catalytic oxidation, can reach tens of degrees.

Fines are deposited in certain locations of the bed, these increasing local resistance and decreasing in turn the local velocity of phases. Lower velocity of the flow further enhances the probability of additional deposits of fines. Structural deviations of this type are typical for catalytic reactors and their effective suppression requires greater mechanical strength and resistance to abrasion of catalytic pellets.

By a similar mechanism a decrease of local void fraction also occurs in packed rectification columns for separation of mixtures that tend to decompose or polymerize with the formation of solid crusts. Induced deviations from the uniform flow, particularly that of liquid, become manifest in the rectification column by a marked decrease of the separating efficiency, this loss being particularly severe if a high degree of separation is required.

d) Deviations from the homogeneity of the structure of the bed in the region of contact of zones of different particles are typical for the operation of blast furnaces. Typical charges of blast furnaces have a distinctly structuralized character due to the segregated charging of the iron bearing components (generally small particles) and coke (generally large particles with a different shape factor). The principal advantage of structuralization of the layer is a decrease of the total aerodynamic resistance of the layer in comparison with the layer charged by mixtures of both components. Under structuralized (segregated) charging the layer in individual zones occupied by particles of the same kind (ore or coke) exhibits a random character with, however, significant deviations in the region of contact between these zones. Mingling of particles in contact zones leads to a decreased area void fraction which is the stronger the greater the disparity of the size of particles of the two components.

2.2. MATHEMATICAL MODELS OF STRUCTURAL INHOMOGENEITY OF THE BED NEAR CONFINING SURFACES

2.2.1. Bed confined by a plane

The mere visual observation through the glass walls of a container of a bed of spheres resting on a plane support reveals that the effect of the support extends for the depth of several sphere diameters from the wall. If we inspect a bed of equal diameter spheres we find that the spheres nearest to the plane support appear with their centers exactly in the distance equal to the radius of the sphere, r. The evidence of the arrangement of the spheres into "strata" becomes less and less apparent with increasing distance from the supporting plane until it completely vanishes in the random configuration of the bed.

According to the concept of the "stratified" model it is assumed that the bed may be divided along its total depth by horizontal planes parallel with the supporting plane while each layer (stratum) contains the same number, N, of spheres. The situation is schematically shown in Fig.2-1. The total thickness of the first layer (stratum) nearest to the supporting plane is $(r + \delta)$; this layer being divided by a plane $i = 1$ at the distance r from the support into two unequal parts.

The plane $i = 1$ is a geometrical locus of centers of all N spheres forming the layer $i = 1$. Fig. 2-1 shows some of the spheres of the first layer designated by letters A, B, C. Depending on the compactness of the bed the spheres may be either in contact, as e.g. spheres A and B, or mutually separated, as e.g. spheres B and C.

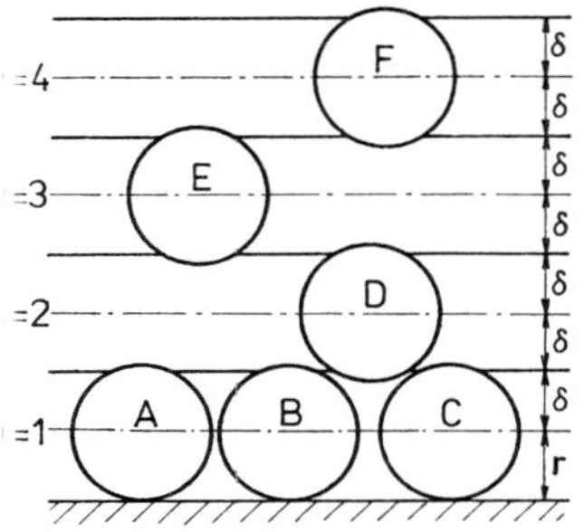

Fig.2-1
Scheme of a layer of spheres confined by a plane surface. The bottom shaded area represents the confining surface. Only few spheres are shown.

On top of the first basic layer there are other layers (strata), each now of thickness 2δ, divided by hypothetical planes, designated by indices $i = 2,3,4,...$, into two equal halves. Corresponding planes are thus planes of symmetry and, at the same time, geometrical loci of the most probable position of centers of spheres belonging to this layer. The probability density of the distribution of the centers of spheres is symmetrical with respect to the plane i and given below by Eq. (2-4).

Some of the spheres with the most probable position of the center within the strata $i = 2,3,4$ are designated in Fig. 2-1 by letters D, E, F. The limits of each stratum, apart from the principal one ($i = 1$) are thus planes, in dependence on the index of the layer i, at the distance $(r + (2i - 3)\,\delta)$ and $(r + (2i - 1)\,\delta)$ from the supporting plane. The maximum vertical distance of two spheres belonging to the same stratum is thus 2δ.

The distribution function of the probability density for the position of the centers of spheres in the i – th layer, $f^i\,(x^i)$, where x^i is the coordinate of the distance of the center of the sphere from the plane i, is taken in the form:

$$f^i\,(x^i) = \frac{1}{2\delta} + \frac{1}{\delta}\sum_n \cos\,(\pi n x^i/\delta)\,\exp\,[-\,(\pi n)^2\,Dh_i/\delta^2] \tag{2-4}$$

where h_i designates the distance of the plane of the most probable position of the centers of spheres in the stratum, i, from the plane $i = 1$.

The function $f^i(x^i) \equiv f^i$ satisfies the following differential equation

$$\frac{\partial f^i}{\partial h_i} = D\,\frac{\partial^2 f^i}{\partial (x^i)^2} \tag{2-5}$$

typical for random diffusion processes whose solution in an infinite region leads to the normal distribution function.

The distribution function (2-4) is valid on an interval $-\delta \leq x^i \leq \delta$ and satisfies the condition of symmetry with respect to the plane i

$$\frac{\partial f^i}{\partial x^i} = 0 \quad \text{for} \quad x^i = 0 \tag{2-6}$$

with the maximum probability density on the plane i. Further it holds that:

$$\int_{-\delta}^{\delta} f^i \, \mathrm{d}x^i = 1 \tag{2-7}$$

Upon expressing the distance h_i in terms of the characteristic length scale of the stratum, δ :

$$h_i = (i - 1)\, 2\delta \tag{2-8}$$

and the quantity D, characterizing the dispersion of spheres around the plane i, in terms of the sphere diameter

$$D = \eta\, 2\, r \tag{2-9}$$

we obtain:

$$f^i\,(x^i) = \frac{1}{2\delta} + \frac{1}{\delta} \sum_n \cos\,(\pi n x^i/\delta)\, \exp\left[-(\pi n)^2\, 4\eta(\mathrm{i} - 1)/\,(\delta/r)\right] \tag{2-10}$$

From the additional properties of this function it follows that for $i = 1$ Eq. (2-10) changes into the Dirac function with a non-zero value at the point $x^i = 0$ which, in fact, is the property of the basic stratum.

In an infinite distance from the supporting plane the distribution of centers of spheres is given by the uniform distribution function

$$\lim_{i \to \infty} f^i = 1/(2\delta) \tag{2-11}$$

This result is identical with the conclusions following for the frequency function of the random layer written in Eq. (2-4).

If the bed of spheres, resting on a supporting plane, is confined in the radial direction by a cylindrical surface of radius R, then the quantity (1-ε) in an arbitrary plane parallel to the support equals the mean square radius of the discs, appearing in this plane by sectioning the spheres, expressed in terms of the radius of the sphere times the dimensionless group $N(r/R)^2$.

Upon designating the dimensionless mean square radius of discs of all spheres sectioned by the given plane as I, the area void fraction in this plane may be expressed by:

$$\varepsilon = 1 - N(r/R)^2 I \tag{2-12}$$

If we introduce for simplicity a dimensionless coordinate of distance, x, from the plane i, related to the dimensional distance, x_0, from plane i by

$$x = x_0^i/\delta \tag{2-13}$$

the dimensionless mean square radius of the discs from all spheres sectioned by a plane in position x satisfying the following inequality

$$1 - (r/\delta) \leq x \leq (r/\delta) - 1 \tag{2-14}$$

may be expressed by

$$\begin{aligned} I &= I^i(x) + I^{i+1}(x) + I^{i-1}(x) = \\ &= 2r/(3\delta) - 4(\delta/r)^2 \sum_n \exp\left[-P(n,i)\right](-1)^n/(\pi n)^2 - \\ &- 2(\delta/r) \sum_n \exp\left[-P(n, i+1)\right] \cos(\pi n(x + (r/\delta)))/(\pi n)^2 - \\ &- 2(\delta/r)^2 (x-1) \sum_n \exp\left[-P(n, i+1)\right](-1)^n/(\pi n)^2 + \\ &+ 2(\delta/r)^2 \sum_n \exp\left[-P(n, i+1)\right] \sin(\pi n(x + (r/\delta)))/(\pi n)^3 + \\ &+ 2(\delta/r)^2 (x+1) \sum_n \exp\left[-P(n, i-1)\right](-1)^n/(\pi n)^2 - \\ &- 2(\delta/r) \sum_n \exp\left[-P(n, i-1)\right] \cos(\pi n(x - (r/\delta)))/(\pi n)^2 - \\ &- 2(\delta/r)^2 \sum_n \exp\left[-P(n, i-1)\right] \sin(\pi n(x - (r/\delta)))/(\pi n)^3 \end{aligned} \tag{2-15}$$

where

$$P(n,i+m) = (\pi n)^2 \eta (i - 1 + m)(r/\delta) \text{ for } m = -1,0,1 \tag{2-16}$$

In regions

$$(r/\delta) - 1 \leq x \leq 1 \tag{2-17}$$

$$-1 \leq x \leq 1 - (r/\delta) \tag{2-18}$$

the expression for the dimensionless mean square radius of discs from all sectioned spheres takes the following form:

$$\begin{aligned} I &= I^{i}(x) + I^{i+1}(x) = \\ &= 2r/(3\delta) \pm 2(\delta/r)^2 (x \mp 1) \sum_n \exp\left[-P(n,i)\right] (-1)^n/(\pi n)^2 - \\ &\quad - 2(\delta/r) \sum_n \exp\left[-P(n,i)\right] \cos\left(\pi n(x \mp (r/\delta))\right)/(\pi n)^2 \mp \\ &\quad \mp 2(\delta/r)^2 \sum_n \exp\left[-P(n,i)\right] \sin\left(\pi n(x \mp (r/\delta))\right)(\pi n)^3 \mp \\ &\quad \mp 2(\delta/r)^2 (x \mp 1) \sum_n \exp\left[-P(n,i \pm 1)\right] (-1)^n/(\pi n)^2 - \\ &\quad - 2(\delta/r) \sum_n \exp\left[-P(n,i + 1)\right] \cos\left(\pi n(x \pm (r/\delta))\right)/(\pi n)^2 \pm \\ &\quad \pm 2(\delta/r)^2 \sum_n \exp\left[-P(n,i \pm 1)\right] \sin\left(\pi n(x \pm (r/\delta))\right)/(\pi n)^3 \end{aligned} \tag{2-19}$$

while in case of doubled plus/minus signs the upper sign holds for the inequality (2-17) and the lower sign for the inequality (2-18).

The expressions (2-15) and (2-19) may be considered to be generally valid for $i > 2$. In the stratum $i = 2$, however, they hold only in region $x \geq (r/\delta) - 1$. In the remaining part of the layer $i = 2$ and the whole layer $i = 1$ there appear in the expressions (2-15) and (2-19) terms relating to the nonexistent stratum $i = 0$. This drawback can be removed by introducing the stratum $i = 0$ with a zero probability density of the distribution of centers of the spheres in the whole stratum:

$$f^0(x^0) = 0 \quad \text{for all } x^0 \tag{2-20}$$

For the purpose of Eqs (2-15) and (2-19) we then broaden the definition (2-16) by:

$$P(n,0) = \infty \tag{2-21}$$

For practical calculations of the profiles of the void fraction in regions affected by the spheres belonging to the first stratum $i = 1$ it is, however, more convenient to utilize special expressions derived with the aid of the Dirac function:

$$f^1(x^1) = \begin{cases} \infty & \text{for} \quad x^1 = 0 \\ 0 & \text{for} \quad x^1 \neq 0 \end{cases} \tag{2-22}$$

instead of the cosine series (2-10).

It remains still to derive an expression for the region directly adhering to the supporting plane. Here we easily find that:

$$I = 1 - \left(\frac{\delta}{r}\right)^2 x^2 \quad \text{for} \quad -\frac{r}{\delta} \leq x \leq 1 - \frac{r}{\delta} \tag{2-23}$$

Upon substituting from the corresponding expressions for the dimensionless mean square radius of the discs, I, into Eq.(2-12) we obtain expressions for the profiles of the void fraction exhibiting typical oscillations near the supporting plane. Details of the derivation of this model may be found in [20].

The formulated stratified model describing the oscillations of the area void fraction near a confing surface contains altogether three parameters. The dimensionless coefficient of damping, η, the dimensionless group $N(r/R)^2$ and the dimensionless length scale (half/thickness) of the stratum, (δ/r).

The parameter η characterizes the rate of damping of the oscillations and the rate of transition from the evidently stratified structure in the proximity of the confing plane to the random structure in sufficient distance from the confining surface. High values of the parameter characterize layers with small depth of "penetration" of the oscillations of the area void fraction and hence rapid transition to the random structure.

In the limit of R tending to infinity and in practice for sufficiently large values of the aspect ratio, (R/r), when the layer may be regarded in the horizontal direction as practically unconfined, the parameter $N(r/R)^2$ becomes independent of the radius of the container, R, because a greater container can accomodate a greater number of spheres, N. Higher values of the dimensionless group $N(r/R)^2$ indicate dense layers with compact configuration of the spheres. Corresponding void fraction profiles are shifted generally to lower values of the area void fraction. In a sufficient distance from the confining plane the layer is again random and the void fraction (in the limit $i = \infty$) reaches a constant value, ε_∞, given, according to the stratified model by

$$\varepsilon_\infty = 1 - N(r/R)^2 \frac{2}{3} \frac{r}{\delta} \tag{2-24}$$

It may be more illustrative to replace the parameter $N(r/R)^2$ by the asymptotic void fraction in region free of the effect of containing surface, ε_∞. Eq. (2-12) thus changes to

$$\varepsilon = 1 - (1 - \varepsilon_\infty) \frac{3}{2} \frac{\delta}{r} I \tag{2-25}$$

Physical meaning of the parameter (δ/r) is clear. It is the dimensionless scale of "wave length" of the oscillations of the void fraction in the proximity of the confining surface. Lower values of (δ/r), however, will cause not only a different wawe length of the oscillations but also lower values of the void fraction in corresponding phase positions.

Kubie [130] analyzed the effect of the plane wall on the distribution of the voidage of a bed of spheres in its immediate proximity. His model starts also from the observation of the highly ordered structure of the bed and assumes that all spheres in the first and the second layer adjacent to the wall are located with their centers on planes parallel to the containing wall at the distance r and $(r\beta_3)$ respectively..

For the void fraction due to the first layer of spheres one can then write

$$\varepsilon = 1 - \beta_1\left[1 - \left(1 - \frac{y'}{r}\right)^2\right] \quad \text{for} \quad 0 \le \frac{y'}{r} \le 2 \tag{2-26}$$

and for the voidage due to the presence of the second layer of spheres then

$$\varepsilon = 1 - \beta_2\left[1 - \left(\beta_3 - \frac{y'}{r}\right)^2\right] \quad \text{for} \quad \beta_3 - 1 \le \frac{y'}{r} \le \beta_3 + 1 \tag{2-27}$$

In region $(\beta_3 - 1) \le (y'/r) \le 2$ where the validity of the two above expressions overlap the resulting void fraction is the sum of the values resulting from the two formulae.

This gives rise to two discontinuities of the slope of the void fraction versus distance profile. Namely, the two parabolas, given by the last two expressions, are interconnected by a shallow segment of another parabola. This is the direct result of the fact that, in contrast to the reality, the centers of the spheres of the second layer are taken to be all on one plane.

The previously presented stratifed model tends to exhibit this type of discontinuity between the individual strata if the values of the coefficient η, characterizing the scatter of the spheres within each strata, are very small.

Kubie [130] recommended the following values for the coefficients of his formulae: $\beta_1 = 0.75$, $\beta_2 = 0.70$, $\beta_3 = 2.7$. For different conditions these values can be determined relatively easily when one realizes that $(1-\beta_1)$ and $(1-\beta_2)$ are minimum void fractions in the first and the second layer. The value of the third coefficient relates to the stratified model by (see Fig.(2-1)):

$$\beta_3 = 1 + 2\delta \tag{2-28}$$

Evaluation of δ using the above given value of the coefficient β_3 gives $\delta = 0.85$, which is in good agreement with its values tabulated later in this chapter in Table 2-1.

An empirical expression for the course of the void fraction near a plane surface as a function of the distance from the plany, y', was proposed by White et. al [94] in the form

$$\varepsilon = \left\{1 + \frac{1 - \varepsilon_\infty}{\varepsilon_\infty}\left[1 - \exp\left(-\frac{y'}{r}\right)\right]^{1/2}\right\}^{-1} \tag{2-29}$$

This expression meets the constraints for the values of the void fraction directly on the wall and at an infinite distance from the plane. In addition, the expression accounts for the existence of the first minimum of the void fraction profile.

Another empirical correlation was proposed by Vafai et. al [93] in the form

$$\varepsilon = \varepsilon_{\infty}\left[1 + C_1 \exp\left(-\frac{y'}{r}\right)\right] \tag{2-30}$$

where C_1 equals unity or 0.9. This correlation provides only for an exponential decrease of the void fraction without oscillations.

Pillai [115] developed a model of a two dimensional packed bed confined by a plane that can also predict voidage oscillations near the containing plane. The idea of the model is that the bed is divided by planes parallel to the confining surface that divides the bed, in the two dimensional case, into strips of equal width such that the radius of spheres forming the bed can be expressed in multiples of the width of the strips.

Based on the geometrical considerations the author [115] then computes the fractional volume of the strip occupied by the spheres with their centers in the investigated strip and the adjacent strips on both sides. The parameters that have to be evaluated from experimental data are the number of the spheres with their centers in the strip, individually for each strip. The author evaluated these numbers from experiments on a two dimensional bed filled with wooden discs. The number of the discs in each strip parallel to the confining surface was counted by visual observation through perspex walls of the bed.

This model, although very instructive, is rather awkward to apply to real three dimensional beds confined by a plane. Nevertheless, the same idea was later used in connection with beds confined by cylindrical surfaces by Govindaro and Froment [116] and the model is explained in more detail in connection with cylindrical beds in the next paragraph.

2.2.2. Bed confined by a cylindrical surface

Govindaro and Froment [116] utilized the same concept as Pillai [115] to develop a mathematical model of the voidage profiles in a cylindrical bed of uniformly sized spheres near the containing wall.

The model is based on the idea of dividing the bed into cylindrical concentric layers (CCL) of equal thickness starting from the wall onwards, see Fig. 2-2. The number of the CCL's, not necessarily covering the whole volume of the bed, is q, while the thickness, or the width of the CCL, is such that the radius of spheres, r, forming the bed is an integral multiple, m, of the width of the layer.

The total volume of the solids (spheres) in a CCL is the sum of the contributions of the caps and slices falling into that particular CCL, see Fig. 2-2. The centers of the contributing spheres do not lie all on the same cylindrical surface concentric to the CCL and therefore an integral average of the volumes of the caps and slices is taken. For simplicity the curvature of the cylindrical surfaces sectioning the spheres is neglected and the caps and slices thus become single-based and double-based segments of a sphere with plane bases.

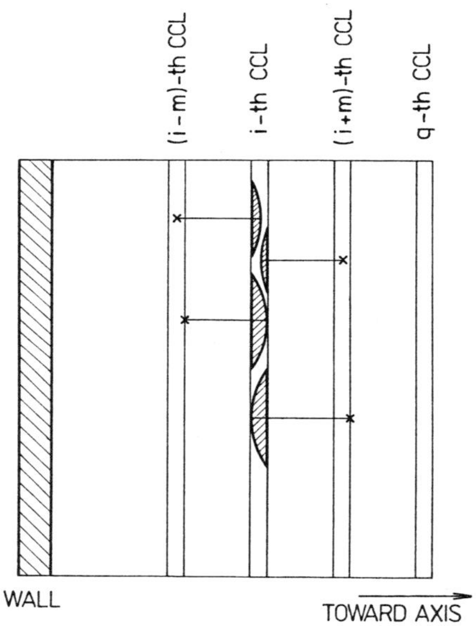

Fig.2-2
Sketch of the Cylindrical Concentric Layers model of the bed confined by a cylindrical surface.

This procedure leads to two explicit expressions for the void fraction. One for the 2m CCL's immediately adherring to the wall, i.e. a region one sphere diameter wide in the immediate vicinity of the wall

$$\varepsilon_i = 1 - \frac{h}{g_i}\left[n_{i+m}\left(m - \frac{1}{4}\right) + 3\sum_{j=m+1}^{i+m} n_j b_{i,j}\right] \tag{2-31}$$

for $1 < i \leq 2m$
and the other for the CCL's farther away from the wall

$$\varepsilon_i = 1 - \frac{h}{g_i}\left[(n_{i-m} + n_{i+m})\left(m - \frac{1}{4}\right) + 3\sum_{j=i-m}^{i+m} n_j b_{i,j}\right] \tag{2-32}$$

for $i > 2m$.

where

$$h = \frac{N_T\, r}{3\, L\, m} \tag{2-33}$$

and

$$g_i = 2\, m \frac{R}{r} - 2i + 1 \tag{2-34}$$

and

$$b_{i,j} = m^2 - i^2 + j(2i - j) - 1/6 \tag{2-35}$$

We note again that in the Govindaro and Froment interpretation, these expressions are valid only for i much smaller than $(m\, R/r)$ due to the neglected curvature of the CCL's.

Provided that the total number of particles in the bed, N_T, the depth of the bed, L, and the geometry of the bed R,r, are known, the only parameters to be obtained from experiments are the number fractions of the spheres in individual CCL's, n_i. These data can be obtained from measurements of the void fraction profiles in CCL's of the dimensions fixed by the integer m.

However, the other parameters are constrained by the following expressions

$$\frac{N_T}{L} = \frac{3}{4\, m^2 r} \left[\sum_{i=1}^{q} (1 - \varepsilon_i)\, g_i + (1 - \varepsilon_\infty) \left(\frac{R}{r} m - q \right)^2 \right] \tag{2-36}$$

through the balance of the void volume in all q CCL's. It was assumed that in the undisturbed bulk of the bed the void fraction equals ε_∞. The last constraint is valid for large values of the aspect ratio R/r permiting the existence of the undisturbed bulk when $q << R/r$. On the other hand, the last equation is inapplicable for an infinite aspect ratio.

Govindaro and Froment [116] found that the choice of the number of CCL's, q, does not really appreciably affect the estimated number fractions, n_i. The highest number fraction fell always into the $(m+1)$-th CCL with the next highest fraction number in the 3m-th CCL at the distance 1.5 d away from the wall. The number fractions in the intermediate CCL's between (m+1) and 3m were zero. This finding is in accord with the stratified model of the layer confined by a plane presented in the previous paragraph where, as can be seen from Fig.2-2, centers of all spheres fall on the plane at the distance of one sphere radius r from the plane. This surface in the cylindrical concentric layers model coincides with the outer surface of the $(m + 1)$-th cell.

The third maximum of occurrence of the spheres in the histogram of the number fractions was detected between $2d$ and 2.35 d, i.e. extending over 2 (for $m = 3$) or over 3 (for $m = 4$) CCL's.

This means that a) the distance between the second and the third layer of spheres was shorter, and, b) the scatter of sphere centers was higher.

Both these conclusions are again in accord with the stratified model of the bed. Equally, the finding that the aspect ratio R/r affected only the amplitude and not the phase of the void fraction oscillations is in accord with the stratified model.

Govindaro and Froment gave also correlations [116] for the number fractions in the $(m + 1)$-th and the $3m$-th CCL's which are really necessary for the calculation of the void fraction profiles from the above equations, but the agreement of the computed profiles based on these correlations with experimental profiles was satisfactory only to the distance of about $2d$ away from the wall.

In a more recent paper Delmas and Froment [109] extended these correlations to account for the decrease of the area cross section of the CCL's from the wall toward the center and the corresponding effect on the number fractions. For $m = 3$ these correlations took the form:

$$n_{m+1} = n_4 = 5.25\left[1 - 1.25\left(\frac{r}{R}\right)^2\right]\bar{n}_4 \tag{2-37}$$

$$n_{3m} = n_9 = 4.65\left[1 + \left(\frac{r}{R}\right)^2\right]\bar{n}_9 \tag{2-38}$$

$$n_j = 2.1\left[1 + 7.4\left(\frac{r}{R}\right)^2\right]\bar{n}_j \quad \text{for} \quad j = 13,\ 14 \tag{2-39}$$

where $\bar{n}_j$ with overbar designate the number fractions of spheres having centers in the j-th CCL in a perfectly random bed, i.e.:

$$\bar{n}_j = \left[1 - (j-1)\frac{r}{Rm}\right]^2 - \left[1 - j\frac{r}{Rm}\right]^2 \tag{2-40}$$

Ridgway and Tarbuck [126] devised a model of the effect of the containing surface starting from a regular close-packed array of spheres. The model is based again on the idea that the layer of spheres nearest to the wall, or the containing surface in general, tends to be highly ordered, while the next layer builds upon the previous one in a less ordered fashion and so on until a fully randomized arrangement is achieved.

For this structure one can derive, based on geometrical argument, expressions for the volume occupied by a single sphere. In order to characterize random beds of spheres these authors introduced three randomizing factors.

The expression for a bed of spheres contained by a cylindrical surface has the following form

$$\varepsilon = 1 - \sum_{0}^{i} \frac{F_1}{2\sqrt{3}} \left(\frac{y'}{r} - \sqrt{\frac{2}{3}}\, iF_2 - \Delta F_2 \right) \cdot$$

$$\cdot \left(2 - \frac{y'}{r} + 2\sqrt{\frac{2}{3}}\, iF_2 + \Delta F_2 \right) \qquad (2\text{-}41)$$

where the bracketed terms must be positive or zero. In the opposite case they are taken to be zero. "i" in this case again is the layer number starting from zero for the first of the spheres nearest to the wall.

The first randomizing factor, F_1, accounts for the fact that the volume occupied by any sphere in a random bed will be larger than in the close-packed regular arrangement.

The second randomizing factor, F_2, was introduced in order to allow for the fact that in a random bed each layer of spheres is a little more opened and permits the spheres from the neighbouring layer to be a little bit closer. From the way both these randomizing factors were introduced it is clear that their value is less than unity and will decrease with increasing distance from the wall, i.e. with increasing layer number i.

The third randomizing factor, ΔF_2, is specific for cylindrically confined beds as for these beds the minima on the oscillating profiles of the void fraction will be somewhat farther away from the wall than in the corresponding bed confined by a flat plane surface.

Putting ΔF_2 equal zero reduces the model in Eq.(2-41) to one for the bed confined by a plane surface. Ridgway and Tarbuck [126] evaluated the first two randomizing factors for this case using data of Benenati and Brosilow [117]. The first randomizing factor is related to the void fraction in each minimum, ε_{min}, on the void fraction oscillation profile by

$$F_1 = \frac{(1 - \varepsilon_{min})}{\pi/(2\sqrt{3})} \qquad (2\text{-}42)$$

where the numerator expresses the actual fractional packing density in the bed and the denominator fractional packing density of the closed-packed hexagonal array. This expression can be easily used to evaluate the first randomizing factor from experimental data. From Benenati and Brosilow [117] data Ridgway and Tarbuck [126] derived the following correlation

$$F_1 = \frac{2\sqrt{3}}{\pi} (0.62 + 0.18^{-0.36i}) \qquad (2\text{-}43)$$

From this result one can easily see that the void fraction of an undisturbed bed (i tending to infinity) was $1 - 0.62 = 0.38$.

For the second randomizing factor Ridgway and Tarbuck [126] found from the same experimental data [117] the following correlation

$$F_2 = 0.991^i \tag{2-44}$$

The third randomizing factor, typical for cylindrical confined beds, was found by Ridgway and Tarbuck, using their own experimental data to exhibit a rather complicated behaviour. For the first layer ($i = 0$) DF equalled 0.1. This value then rose to about 0.175 for the third ($i = 2$) layer and then dropped gradually to zero for the fifth and the following layers where the effect of curvature of the wall was offset by increased randomness. These results were obtained for 9 mm spheres in a 155.6 mm in diameter container and are expected to be a function of the aspect ratio R/r.

Martin [88] formulated a model of the void fraction profiles near a cylindrical surface which can be classified as a trigonometric model as it utilizes trigonometric function to describe the void fraction oscillations near the wall. The model is defined by means of the coordinate defining the radial distance from the wall, y', in terms of particle radius, r, while the zero is at the point of the first void fraction minimum:

$$y = \frac{y'}{r} - 1 \tag{2-45}$$

The void fraction profile within one sphere radius from the wall is then given by geometry considerations as

$$\varepsilon = \varepsilon_{min} + (1 - \varepsilon_{\mathrm{min}})\, y^2 \quad \text{for} \quad -1 \leq y \leq 0 \tag{2-46}$$

where $\varepsilon_{\mathrm{min}}$ is the void fraction corresponding to the first minimum on the profile and ranges usually between 0.2 and 0.26.

For positive values of the coordinate y, i.e. distances more than one sphere radius away from the wall, the void fraction is given by the following empirical formula

$$\varepsilon = \varepsilon_\infty + (\varepsilon_{\mathrm{min}} - \varepsilon_\infty) \exp\left[-y/4\right] \cos\left(\pi\, y/T\right) \tag{2-47}$$

The "wave length", T, of the void fraction oscillations, is measured also as a fraction of particle size.

For the values of the void fractions $\varepsilon_{\mathrm{min}} = 0.23$ and $\varepsilon = 0.39$ and infinite aspect ratio R/r, T equals $\sqrt{(2/3)} = 0.816$, while $T = 0.876$ for $\mathrm{R/r} = 20.3$ based on data of Benenati and Brosilow [117].

Martin [88] further correctly noted that in pseudohomogeneous models one should remain on the scale of particle size and found that in channels between two adjacent void fraction minima the average void fraction varies little and equals practically the void fraction ε_∞. The bed thus can be divided into two zones: One adhering to the wall one particle radius wide, where the mean void fraction ε_w, obtained by integration of Eq. (2-46) in the limits of y between -1 and 0, is

$$\varepsilon_\infty = 1 - \frac{2}{3}\left(\frac{R/r - 7/8}{R/r - 1/2}\right)(1 - \varepsilon_\infty) \tag{2-48}$$

The rest of the bed between $z = 0$ and $((R/r)-1)$ has the bulk void fraction ε_∞. In this form the model falls into the category of zone models. It is interesting to note that the void fraction for the wall region, given by Eq.(2-48), varies relatively little with the aspect ratio R/r.

Another empirical model of the void fraction profiles that can be classified as a trigonometric one was formulated by Chandrasekhara and Vortmeyer [83] in the following form

$$\varepsilon = \varepsilon_\infty\left(1 + C_1 \exp\left[-C_2\frac{y'}{d}\right]\right) \tag{2-49}$$

Numerical values of the two empirical constants were adjusted on the basis of data of Benenati and Brosilow [117] to $C_1 = 1$ and $C = 3$ for the aspect ratios R/r equal 14 and 21.

In a later paper Vortmeyer and Schuster [92] relaxed the requirement of two adjustable constants to a single constant:

$$\varepsilon = \varepsilon_\infty\left(1 + C_3 \exp\left[1 - \frac{y'}{r}\right]\right) \tag{2-50}$$

The latter equation was used both for rectangular and cylindrical beds and C_3 has to be adjusted according to ε_∞ .

Cheng and Hsu [114], in fact, used Eq. (2-49) to describe the void fraction profile in an annular bed near both of its walls, i.e. near the convex and concave confining surface. The value of the parameter C_2 in this case was set equal to 2 which reduces Eq.(2-49) effectively to Eq.(2-50).

The parameter C_1 has been so far always put equal to unity. This results in the void fraction immediately near the wall being not equal to unity as observed on the experimental void fraction profiles. The reason frequently put forward for not respecting this experimental observation is that the corresponding velocity at the wall, computed from equations that do not account for the friction on the wall, should tend to infinity.

However, severe difficulties and convergence problems are encountered if one uses equations allowing for the wall friction with the void fraction equal to unity on the wall. For this reason several authors compromized [95,83,114] by adopting values of the void fraction near the wall of less than unity.

Into the class of the trigonometric models of the void fraction oscillations falls also the model of Cohen and Metzner [91], basically an empirical model whose constants are typical for a bed of spheres and for a range of aspect ratios

$$\frac{1-\varepsilon}{1-\varepsilon_\infty} = 4.5\left[\frac{y}{d} - \frac{7}{9}\left(\frac{y}{d}\right)^2\right] \quad \text{for} \quad \frac{y'}{d} \leq 0.25 \tag{2-51}$$

$$\frac{1-\varepsilon}{1-\varepsilon_\infty} = 0.3463 \exp\left[-0.4273\frac{y'}{d}\right]\cos\left[\pi\left(2.4509\frac{y}{d} - 2.201\right)\right]$$

$$\text{for} \quad 0.25 < \frac{y'}{d} < 8 \tag{2-52}$$

$$\varepsilon = \varepsilon_\infty \quad \text{for} \quad 8 \leq \frac{y'}{d} \leq \infty \tag{2-53}$$

From Eq.(2-52) one can evaluate the "wavelength" of the void fraction oscillations to be 0.780 of particle diameter which is in reasonable agreement with the data shown in Table 2-2 in connection with the stratified model and low aspect ratios.

However, for calculations of the velocity profiles in beds of spheres Cohen and Metzner [91] defined a triregional model consisting of the wall region (one particle diameter wide and adhering to the wall), the transition region (adjacent to the wall region up to five particle diameters away from the wall) and the bulk region.

In the bulk region, the void fraction is that of an undisturbed bed and given by Eq.(2-53). In the transition region the local void fractions are given by Eqs(2-51) and (2-52), while in the wall region the void fraction is given by Eq. (2-51), but account is made in the calculation of the hydraulic radius of the wetted surface of the wall. Also from the trizonal model it can be seen that a bed 10 particle diameters in diameter has no undisturbed bulk.

The zone model of Martin [88], as well as the empirical models of Chandrasekhara and Vortmeyer [83] and Vortmeyer and Schuster [92], do not reflect the void fraction oscillations. The justification for these models stems from the fact that on the macroscopic scale the oscillations of the void fraction tend to disappear. This has been shown by Benenati and Brosilow [117] in the form of a plot of the integrated void fraction (by integrating the local void fraction over the distance from the wall) which showed only minor oscillations. In addition, Schuster and Vortmeyer [118] found that imperfections of the shape and surface of the spheres forming the bed bring about significant damping of the void fraction oscillations.

2.3. EXPERIMENTAL DETERMINATION OF INHOMOGENEITIES IN FIXED BEDS

In the majority of studies the experimental procedure of the void fraction measurement was based on fixing the bed by a resin or wax. Probably the first authors to use the wax were Roblee et al. [123]. Epoxy resin was used by Benenati and Brosilow [117], Thadani and Peebles [119], and Propster and Szekely [121] and others. The structure of the bed fixed by the resin is then analyzed by machining away thin layers of the bed and weighing the removed material. The void fraction in the removed material is easily calculated from the material balance and the known differences in the density of the resin and the packing material.

An advantage of this method is that it permits measurement of the void fraction profiles both in cylindrical layers removed on the lathe and in the plane horizontal slices. Propster and Szekely [121] employed the method to investigate the void fraction profiles also in the "interfacial" region between layers of particles of different size.

A certain disadvantage of the fixing method is the destruction of the bed and definite limitations put on the choice of the packing material.

Thadani and Peebles [119] also tested an X-ray or gamma-ray absorption method utilizing different absorptivity of the packing and the matrix of the material fixing the bed. Gamma-ray absorption to measure the void fraction in coke and ore particle beds was also used by Schultz and Abel [122]. The source of gamma-rays was Co^{60}. Upon passing through the bed the narrow beam was detected by the Geiger-Mueller counter. Extinction of the beam depends, among others, on the material, its density and thickness. Consequently, the apparatus has to be calibrated for each particle material and measurement of mixtures of particles of different materials is difficult.

Ridgway and Tarbuck [48] employed a technique based on rapid rotation of a cylindrical bed to which a measured volume of liquid is added. The void fraction is then computed from an increase of the liquid level pushed to the wall by the centrifugal force.

An interesting optical method was developed by Kalny [120]. The bed is made up of glass particles that practically do not absorb light. The voids are filled by liquid, a benzene-ethanol or benzene-chlorbenzene mixture, coloured by a dye. The void fraction was obtained from measurement of the absorption of a monochromatic light beam passing through the bed in various positions of the column cross section. The important condition for the success of the method is optical isotropy of the bed. This was accomplished by adjusting the concentration of the two-component liquid mixture to the same refractive index as that of the glass of the packing.

Schultz and Vortmeyer [118] developed a rapid technique of measuring the void fraction profiles consisting of removing a layer of particles by a surface coated by a glue. Each layer removed was photographed and the results were obtained in the form of a histogram for classes a particle diameter wide.

In order to evaluate parameters of a two dimensional model of the void fraction profiles Pillai [115] used wooden discs as particles. In this case the void fraction can be easily computed from visual observation (counting) of the bed through perspex or glass walls.

The technique employed by the author of this book determines what is approximately the area void fraction from an increase of the liquid level in the bed following addition of a known volume of liquid. The liquid is added through the hole in the bottom of the bed and the increase of the liquid level is measured in a parallel tube by means of a cathetometer [22] or by a sensitive pressure transducer. This technique permits measurement of the area of void fractions in horizontal layers. The volume of liquid (ethanol) added is adjusted so as to obtain the area void fractions with sufficient accuracy. For too small volumes of liquid added the relative error of determination of the increase of the liquid level, which is decisive, is too large. If the volume added is too large, the data lose the character of area void fraction.

An advantage of the described method is that it is nondestructive and does not pose any limitations on the choice of the packing material. It was used to measure void fraction profiles near a confining plane and in the "interfacial" region between layers of particles of different diameter.

A method based on equalizing the refractive index of particles (glass) and liquid was developed also by Schneider and Rippin [153]. These authors used mixtures of toluene and benzyl alcohol to achieve optical homogeneity of the bed composed of coloured glass particles.

The bed was lit by monochromatic light to minimize scatter. Alternatively, the authors used also a methyl red dye to colour the solution. The bed can then be photographed, or, as the authors suggested, projected using an overhead projector, while retaining excellent definition, brightness and resolution of the image. The images were in turn evaluated by a photocell and the results were thus values at various points of the column cross section. Schneider and Rippin [153] themselves recommended direct absorption measurement on the experimental equipment with a photocell positioned directly below the monochromator.

There are definite advantages of the methods based on equalization of the refractive index of the packing and the liquid filling the bed. Firstly, the method is nondestructive and fast, permits measurement at various points of the bed cross section. The measurement is not limited to beds exhibiting certain symmetry (cylindrical beds) and local measurements can be made, as shown by Schneider and Rippin [153], in beds with cooling tubes.

A certain disadvantage of the method is that it is confined to transparent particles and also the depth of the bed is limited. Schneider and Rippin [153] put this limit at about 10 particle diameters.

An attractive alternative to experimental determination of the void fraction is the use of computer simulations [125,135- 140,154]. Such simulations are based on observing a cluster of particles formed by adding one particle at a time to the cluster while observing geometrical constraints for the position of the particle. A different approach simulates the actual formation of the bed by dropping individual particles into the bed while the particle upon reaching the bed is allowed to attain a stable position.

Computer simulations are easily amenable to studying the effect of the containing surfaces [125], void fraction distribution in complex particle mixtures (size distribution) and for particles of complicated shape (e.g. significant non-sphericity).

Dickinson et al. [154], for instance, found by a two dimensional computer simulation of spheres (in two dimensions actually discs) that minor deviations from equal sized particles to slight polydispersity increases the void fraction while additional, more significant increases of polydispersity have a much less marked effect.

Beds with minor polydispersity tend to exhibit much smaller fluctuations of local average void fraction and much more uniform void fraction than monodisperse systems. Approximately the same was found for the pore size distribution. In this case, however, as the degree of polydispersity increases there is an increasing number of large pores present in the bed.

2.4. AREA VOID FRACTION IN A FIXED BED CONFINED BY A PLANE

A large number of experimental measurements of the area void fraction in the bed confined by a plane [22] were carried out using the method of detecting the rise of liquid level. The confining surface in this case was represented by a flat bottom of a cylindrical vessel of internal diameter 247 mm.

A typical profile of the area void fraction in dependence on the distance from the bottom is shown in Fig.2-3 for spheres 10 mm in diameter. The curve in this figure was computed from the stratified model, Eq.(2-12). Optimum values of all three parameters involved were determined by nonlinear regression using the Marquardt algorithm [68].

From the extensive set of analogous measurements one can draw the following conclusions:

The agreement of the model with the experimental profiles of the area void fraction is consistently good. Systematic deviations appear only in the region of the first minimum. This may be apparent also from Fig.2-3. It thus appears that the real number, N, of the spheres in the first hypothetical stratum adhering to the confining plane (bottom) is somewhat higher than in the remaining strata. This explanation is further supported by the observation that the optimum values of N computed from the model are in all cases somewhat lower than the maximum number of spheres that can be placed on the bottom of the measuring container. This maximum number of spheres on the bottom was determined by an auxiliary experiment.

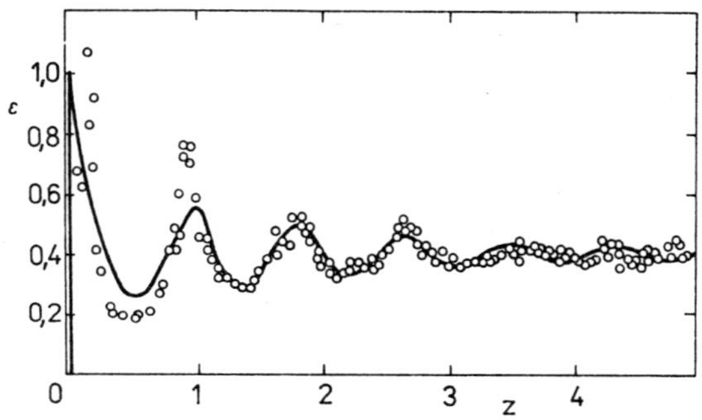

Fig. 2-3
Profile of area void fraction in a bed of 10 mm spheres confined by a plane.

For 15 mm spheres and smaller the values of the dimensionless half thickness (δ/r) are identical within experimental error, i.e. independent of sphere radius, r, and, moreover, independent of the method of charging the bed. Numerical values are summarized in Table 2-1. This table gives also brief characteristics of the two methods of charging the bed by spheres. The tabulated data agree well with the value (0.85) of (δ/r) that can be estimated from the void fraction profile published by Benenati and Brosilow [48].

Within experimental error the values of the parameter η, characterizing the rate of damping of the void fraction oscillations, do not depend on the sphere radius either. Numerical values of this parameter are also shown in Table 2-1. They indicate that rough methods of charging, such as dumping the spheres into the empty container, lead to lower values of the parameter and hence to deeper penetration of the oscillations.

Table 2-1

Characteristic values of parameters of the stratified model for beds with negligible effect of container walls

Method of charging	(δ/r)	η	$N(r/R)^2$	ε
Dumping particles into empty container	0.843	0.00542	0.780	0.383
Dumping particles into container filled with ethanol	0.842	0.00948	0.752	0.405

Numerical values of the parameter $N(r/R)^2$, for sufficiently small ratios, (r/R), are independent of this ratio. Since for spheres 5 – 15 mm in diameter, investigated experimentally in [22], the effect of the container walls could be regarded as negligible; the values of $N(r/R)^2$ no longer depend, for the given container, on the sphere radius, r. Characterist . values of $N(r/R)^2$ are given in Table 2-1, together with the corresponding computed asymptotic void fraction ε_∞ . The stratified model itself, however, has only three parameters while $N(r/R)^2$ and ε_∞ are two alternatives for only one of them, depending on whether one uses Eq. (2-12) or (2-25). From the tabulated data it follows that their dependence on the method of charging the vessel and the conclusion that more gentle methods of filling, such as e.g. dumping the spheres into the container filled by ethanol, lead to less compact beds with a lower characteristic number N and, on the contrary, higher asymptotic void fraction ε .

The above given conclusions, however, do not hold for the results obtained [22] with 20 mm spheres. In this case the aspect ratio equals $(R/r) = 12.35$ and the bed is markedly inhomogeneous in the whole volume due also to the effects of the container walls (see Table 2-2). For geometry reasons the configuration of spheres in the container cannot be sufficiently compact and individual hypothetical strata of the stratified model therefore interfere to a greater extent. This results in lower values of the half thickness (δ/r). In case of dumping the spheres into the container filled by ethanol the lower number of spheres in a single stratum is, however, more than compensated by closer configuration of the strata and the result is a lower asymptotic void fraction (see Table 2-2). As further follows from Table 2-2 the half thickness of the stratum is also essentially independent of the method of charging the bed.

Table 2-2

Characteristic values of parameters of the stratified model under the effect of container walls $(R/r) = 12.35$

Method of charging	(δ/r)	$N(r/R)^2$	ε_∞
Dumping particles into empty container	0.793	0.724	0.391
Dumping particles into container filled with ethanol	0.796	0.732	0.387

Numerical values of the parameter characterizing the damping of the void fraction oscillations reflect the fact that the oscillations are marked up to about 4 to 5 particle diameters from the confining surface, see also Fig. 2-3. From this it follows that beds with the ratio R/r approaching or even equal to 10 are inhomogeneous in the whole volume as the effect of the cylindrical containing walls is, as will be seen subsequently, very similar to that of the plane confining surface.

This statement may be supported also by the findings of Ridgway and Tarbuck [48] who found experimentally periodic oscillations of the area void fraction indicating the dimensionless wave length of these oscillations to be 0.84, which is a value identical with the optimum value given in Table 2-1. Similar conclusions have been arrived at by Kondelík, Horák and Tesařová [49].

Although on the top the bed is not confined spatially by a plane the oscillations appear here as well as may be apparent from Fig.2-4. This figure shows the course of the area void fraction as a function of the vertical distance from the top expressed in multiples of particle diameter. The oscillations are here also clearly visible and appear to be due to the leveling of the top of the bed. The curve in Fig.2-4 was also computed from the stratified model with optimized parameters. The deviations in the region of the first minimum suggest that in the absence of the confining surface the spheres in the first hypothetical stratum need not be exactly one sphere radius away from the "confining" surface. Also the opposite pull of gravity, away from the confing plane tends to loosen the imposed regularity of arrangement. The damping of the oscillations in this case is therefore faster and the oscillations decay at the distance of about 2 to 3 particle diameters.

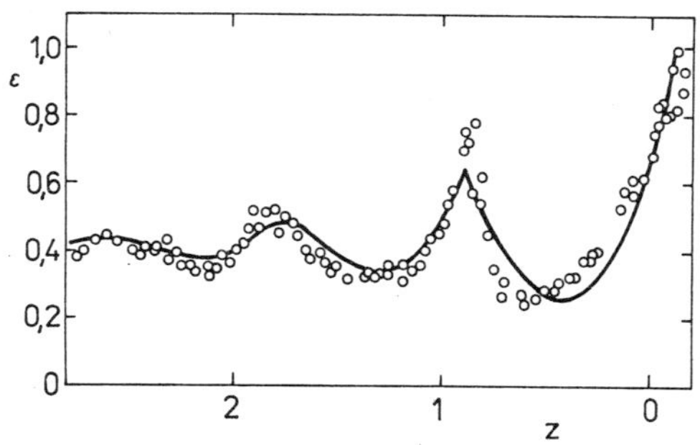

Fig. 2-4
Profile of area void fraction in a bed of 15 mm spheres near its leveled top.

An analogous phenomenon with far reaching consequences for the homogeneity of the bed appears if the container is filled with particles gradually in doses while leveling each dose in the container by compressing it by a plane surface. Characteristic oscillations then appear in each dose following the same mechanism as that near the top of the bed. The resulting bed then exhibits oscillations of the void fraction along the vertical coordinate in the whole volume. This observation was made experimentally also in the paper [22].

Gotoh et al. [125] studied the effect of the confining plane surface on the void fraction of a randomly packed bed of spheres using computer simulation. They noted that the void fraction variation near the plane wall is almost the same as that around a central sphere in the interior of the bed. One particle can be assumed to have an infinite radius so that the problem becomes equivalent to one of a plane surface. The random packing was then regarded as a model of an instantaneous molecular configuration of simple fluids. The probability of finding a sphere center at a given distance from the surface (wall or sphere) was taken to be an analytical solution of the Percus-Yevick approximation of the radial distribution function. The computed pattern of the void fraction profiles also displayed oscillations and it was observed that when the bulk mean void fraction of the bed decreases (more compact beds) the particle arrangement near the confining plane becomes more ordered and the peaks on the oscillatory profile have a greater "amplitude".

2.5. AREA VOID FRACTION IN A FIXED BED CONFINED BY A CYLINDRICAL SURFACE

In general the experimental void fraction profiles in a bed of equal spheres near a containing cylindrical surface can be described as follows. The void fraction starts at unity on the cylindrical surface and reaches a minimum of about 0.25 about a sphere radius away from the wall. The subsequent part of the void fraction profiles are damped periodic oscillations that die away about 5 particle diameter diameters from the wall.

This picture was essentially drawn by Benenati and Brosilow [117]. The same results have been later arrived at by different experimental techniques by other authors. Propster and Szekely [121] observed this behaviour on beds of 1.3, 3.8, 4.8 and 7.6 mm in diameter lead shot. Thadani and Peebles [119] found also good agreement with Benenati and Brosilow data on a bed of 3/8 inch perspex spheres and Roblee et al. [123] for a bed of 0.49 inch spheres. The same results have been arrived at by Schuster and Vortmeyer [118] for 4 mm steel spheres. In contrast, however, these authors observed a much stronger damping of the oscillations when measuring with slightly imperfect shape glass spheres.

Actually, the void fraction minimum in the first row of spheres adjacent to the wall is not exactly one sphere radius away from the wall. This is the case of only the plane wall as shown before. Near the cylindrical surface the minimum is a little farther away and this distance, x_R^0, can be related [117] to the aspect ratio R/r by

$$\frac{x_R^0}{2r} = \frac{1}{2} + \left(\frac{R}{r} - 1\right) - \sqrt{\left[\left(\frac{R}{r}\right) - 1\right]^2 - 1} \tag{2-54}$$

Roblee et al. [123] carried out also measurements with nonspherical particles such as cylinders, Raschig rings and Berl saddles. For 1/2 inch cylinders they found analogous behaviour as that observed for spheres but for Raschig rings the behaviour was quite different. Initially the void fraction decreased from unity on the confining wall to a minimum about 1/8 of a diameter from the wall. Up to this distance (corresponding to the inner void ring of the Raschig rings) the course of the void fraction was the same as for the cylinders. However, in the subsequent part the void fraction increased from 0.63 to a maximum of 0.73 about 1/4 diameter the wall. Then the void fraction decreased to about 0.63 about 1/2 diameter from the wall and remained essentially constant.

For Berl saddles, according to the observation of the same authors [123], the void fraction decreased from unity at the wall to a relatively constant value one particle diameter away from the wall with no oscillations observed. This was attributed to the high non-sphericity of the shape of Berl saddles.

Kondelík et al. [49] investigated experimentally also cylinders in region of the first minimum and maximum and found that the void fraction of the first minimum was between 0.2 and 0.1, i.e. substantially less than for spheres, while this minimum was farther away from the wall than in case of spheres. Using geometrical considerations they derived expressions for the limits between which the void fraction minimum should appear in dependence on the aspect ratio R/r. All experimental void fraction minima fell within these predicted limits.

Benenati and Brosilow [48] also carried out experiments in a cylindrical bed packed with spheres with a cylindrical rod mounted coaxially within the bed. The void fraction profiles as a function of the distance from the central rod displayed analogous oscillations starting from unity on the surface of the rod. The position of the first minimum, however, was in this case shifted from the position one radius away, typical for the plane surface, closer to the rod. The amplitudes of the following oscillations were damped and died away also about 4 to 5 sphere diameters from the rod.

Schneider and Rippin [153] measured void fractions within the radial flow reactor in the form of a tubular heat exchanger where the catalyst particles filled the space between the cooling tubes. The void fraction around the cooling tubes was found to oscillate just as in the neighbourhood of the wall. A hexagonal regular zone could be identified surrounding each cooling tube where the packing developed undisturbed rings. In the corners of these hexagons stacks of single spheres and other simple geometries could be identified.

2.6. AREA VOID FRACTION IN REGION OF CONTACT OF BEDS OF UNEQUAL PARTICLES

Deviations from the homogeneity of the structure of the bed appear also in region of contact of particles of different size. These inhomogeneities can best be studied also with layers of spheres of different diameter.

The measurements of this type were carried out in [21] in such a way that a cylindrical container was first charged with spheres of a certain diameter and experimental measurements provided profiles of the area void fraction along the vertical coordinate. The bed was then dried by the stream of air and a layer of spheres of different diameter than the former was charged into the container. The measurement of the area void fraction profile along the vertical coordinate was then repeated.

Typical profiles of the area void fraction for the two above described experiments are shown in Fig.2-5. In this case the bottom layer is formed by 20 mm spheres while the top layer by 5 mm spheres. The empty circles are experimental points from the first measurement of the area void fraction when there were only the larger spheres in the container.

Fig. 2-5 shows that the area void fraction is initially approximately constant and begins to increase in the region near the top of the bed until it reaches unity in the horizontal cross section entirely above the bed of particles. Full circles in the same figure depict the course of area void fraction in the container with both layers of spheres. At the far right, i.e. for large values of the vertical coordinate and hence deep in the top layer of smaller spheres, the void fraction is again nearly constant and approximately the same as in the region well within the bed of larger spheres. This is in accord with the fact that the void fraction of a random bed of spheres is independent of their size.

From Fig.2-5 it is further apparent that in part of the region originally occupied by the larger spheres only, the full circles are systematically below the empty ones. The decrease of the area void fraction in this "interfacial" region occurs due to the filling of the interstices in the matrix of the larger spheres by the smaller spheres.

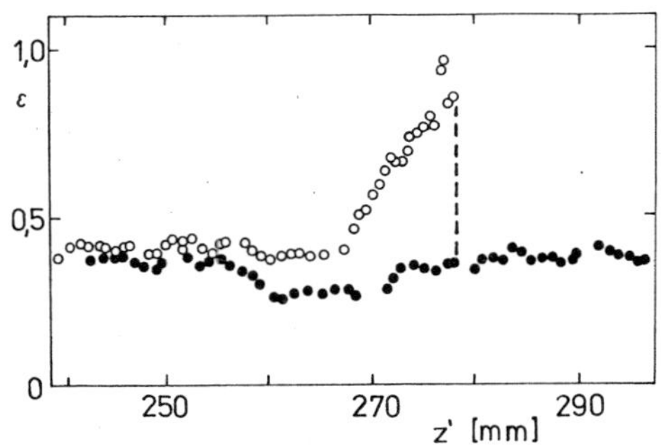

Fig.2-5
Profile of area void fraction in region of contact of 20 and 5 mm sphere zones.
Empty circles - layer of 20 mm spheres only.
Full circles - layer of 5 mm spheres on top of the original 20 mm sphere layer.

If the bottom layer of the larger spheres is slightly compressed after charging into the container by a plane surface, the void fraction oscillations already mentioned appear. The lowering of the void fraction in the interfacial region is then superimposed on these oscillations. This situation is seen from Fig.2-6 again for 20 mm spheres forming the bottom layer and 5 mm spheres forming the top layer. The penetration and mingling of particles in the "interfacial" region is quite obvious from this figure. In comparison with the previous figure, however, the decrease of the void fraction is enhanced particularly in the first minimum of the void fraction which is deeper. Systematic decrease of the void fraciton is evident up to the depth of about 3 diameters of the bottom spheres.

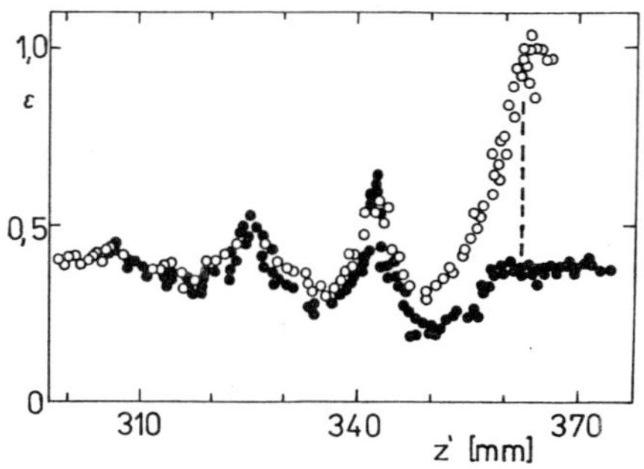

Fig.2-6
Profile of area void fraction in region of contact of 20 and 5 mm sphere zones. Empty circles - - layer of 20 mm spheres only leveled by gentle compression after charging. Full circles - layer of 5 mm spheres on top of the original 20 mm sphere layer.

The results of the extensive set of measurements [21] with spheres may be generalized as follows. In region of contact of layers of particles of unequal diameter, when the bottom layer is made up of the larger particles, a decrease of the area void fraction always occurs in comparison with the values corresponding to the random layer. The extent of the region of decreased void fraction is the larger the greater the disparity of particle diameters forming the two layers. For spheres this disparity may be characterized by the ratio (d_t/d_b), where d_t and d_b designate the diameters of spheres forming the top and the bottom layer respectively. The decrease of the area void fraction in the "interfacial" region appears also if the interface is first leveled by compressing it by a plane surface. Under such circumstances the decrease of the void fraction is superimposed on the oscillatory profile of the void fraction along the vertical coordinate.

The region of decreased area void fraction appears also in the opposite case when the bottom layer is formed by smaller particles provided that the disparity of particle diameters is sufficient. The mechanism of the appearance of the region of decreased void fraction is as follows. The larger spheres, after being dumped onto the layer of the smaller spheres, deform the bottom layer. During this deformation the larger spheres penetrate into the region originally occupied by the smaller spheres while pushing simultaneously the smaller spheres upwards. The region of penetration and mingling of the spheres of different diameter thus forms again.

For small disparities of diameters of the spheres, which for geometry reasons do not permit mutual mingling, changes on the void fraction profiles nevertheless appear particularly in the region of the top layer. The bottom layer now functions as a "plane" support giving rise to the oscillatory course of the area void fraction in dependence on the vertical coordinate.

Analogous measurements of the area void fraction were carried out [28,42] also with samples of particles of coke (10 to 50 mm) and agglomerate (0.5 to 10 mm). These samples substantially did not contain fines nor extremely large particles. Analysis of the results of measurement of the profiles of area void fraction in the region of contact of particles of this type lead to essentially the same conclusions as those yielded by measurements with spheres.

Propster and Szekely [121] also investigated horizontally layered beds of lead shot of diameter ranging between 1.3 and 7.6 mm. Their measurements of the course of the void fraction layers when the top layer was made up of smaller particles showed the existence of sharp void fraction minima which were attributed to the penetration of the smaller particles from the top layer into the bottom layer. The depth of penetration was found to be greater for larger disparities of particle diameters and the depth of penetration amounted to about 2 to 6 bottom (larger) particle diameters, d_b.

In the opposite case when larger particles were deposited on top of the layer of smaller particles the depth of penetration was smaller and amounted to about 1 to 2 top (larger) particle diameters, d_t. In all their results one can observe oscillations of the void fraction in the interfacial region of particle penetration and partly also in the region of the bottom layer. These oscillations can be attributed to compression of the bottom layer of particles intended to level its surface after filling. This is a clear analogy of the results shown in Fig.2-6.

Propster and Szekely [121] constructed a plot of the area void fraction in the interfacial region between layers of spheres, when the smaller particles were placed on top of the larger ones, as a function of the vertical coordinate, expressed in terms of diameters of the larger spheres, d_b. This plot shows that the void fraction minimum grows deeper with the increasing d_b/d_t ratio and drops to about 0.14 for the above ratio equaling 6. At the same time the local minimum is progressively more widely open indicating progressively greater extent of the interfacial region.

Another plot constructed by Propster and Szekely [121] of the extent of the interfacial region, expressed in terms of the bottom sphere diameter, d_b, versus the d_b/d_t ratio, indicated a minimum extent of the interfacial region for the above ratio equaling unity. It should be, of course, theoretically zero as we are dealing with monosized spheres. The extent of the interfacial region then sharply increased on both sides of the minimum. In region of d_b/d_t between 2 and 6 the extent of the interfacial region, expressed in terms of d_b, increased approximately linearly with d_b/d_t.

Approximately it can thus be concluded that the dimensional extent of the interfacial region, when the smaller particles rest on top of the larger ones, is directly proportional to the square of the diameter of the larger sphere, d_b, and indirectly proportional to the diameter of the smaller sphere, d_t.

Schultz and Abel [122] investigated also the void fraction in regions of horizontally layered particles of formed coke, lumped coke, sinter ore and pellets. The size of these particles ranged between about 10 and 100 mm. The charging of the bed followed the blast furnace charging practice and thus the layers of particles produced were sloping from the center toward the walls of their rectangular bed 1.51 × 0.5 × 2.5 m. One of the walls was made of perspex to permit visual observation.

The observed regions of decreased void fraction due to the penetration of particles between coke/ore layers ranged in size between 5 cm up to complete mixing of both layers. Greater penetration was also observed for larger particle disparities. In addition, however, it was observed that sinter ore particles penetrate less than pellets. Sinter penetration into layers of formed coke was observable when the coke to sinter particle diameter reached values about 3 to 4 (estimated from their data). On the other hand, pellets increasingly penetrated the coke layers with increasing coke particle diameter to complete mixing for the particle diameter ratio (estimated) about 8.

Depending on the depth of penetration, minima on the void fraction curve of different width were observed with the extreme value for the minimum void fraction about 0.16. The large extent of the interfacial region, or the depth of penetration, compared to that found by other authors has to be attributed to the charging practice that simulated charging of a blast furnace which certainly promotes penetration in comparison with the method in which one layer is placed on top of another previously prepared and penetration takes place solely by "filtration".

Regions of lowered void fraction, of course, will appear not only in horizontally layered beds or in beds with slanting layers of particles of different diameter. Propster and Szekely [121] investigated also the case when a cylindrical bed is made up of two coaxial regions and the two regions, cylindrical around the axis and annular adjacent to the wall, are packed by particles of different diameter. Such situations can be attractive as a means of affecting the flow distribution in the bed.

Experimental results of the void fraction profiles in the radial direction obtained by the bed fixing technique indicated also an interfacial region of local void fraction minima in the region of contact of particles of different diameter. The extent, however, was smaller, about 1.5 to 2 larger particle diameters. The extent and the depth of the void fraction minimum also depend on the ratio of particle diameters in the central and the wall region. The smaller extent of the interfacial region is directly attributable to the impaired conditions for penetration of the smaller particles into the adjacent matrix of the larger particles when gravity is the major force for the penetration process. The penetration of the smaller particles into the matrix of the larger ones is thus definitely directional as the vertical penetration takes place more readily than the radial one.

List of principal symbols to chapter 2

d	sphere diameter
d_t, d_b	sphere diameter in top and bottom layer
D	coefficient of dimension of length characterizing scatter of spheres about plane i, defined in Eq. (2-5)
f^i	probability distribution function of the positionof centers of spheres in the i-th stratum
F_1, F_2, ΔF_2	randomizing parameters
h_i	distance of the i-th plane from plane $i = 1$, defined in Eq.(2-8)
i	index of sequence of planes of symmetry for individual strata, index of stratum
I^i	dimensionless mean square radius of discs from sectioned spheres belonging to stratum i
I	dimensionless mean square radius of discs from all sectioned spheres, defined by Eqs (2-15) and (2-17)
L	depth of bed
N	number of spheres in a single stratum of the stratified model
N_T	total number of spheres
$P(n,i)$	quantity defined in Eqs (2-16) and (2-21)
r	sphere radius
R	radius of cylindrical container
x	dimensionless coordinate of distance from plane i, defined in Eq. (2-13)
x^i	distance of center of sphere from plane i
x_0^i	coordinate of distance from plane i
y'	distance from surface
y	dimensionless distance from the first void fraction minimum, defined in Eq. (2-45)
z'	vertical coordinate, distance from surface
z	dimensionless vertical coordinate as a multiple of d
δ	half thickness of stratum of the stratified model
η	dimensionless coefficient of dispersion of spheres about plane i, defined in Eq. (2-9)
ε	void fraction
ε_∞	void fraction of perfectly random bed

Additional symbols defined in text.

3. Velocity and pressure field in an inhomogeneous fixed bed

In the previous paragraph we have analyzed various types of structural inhomogeneities of a bed of particles, appearing as undesirable phenomena, or, on the contrary, created purposefully with the aim positively to influence the thermal regime, and, eventually the course of the chemical reaction taking place within the bed.

Subsequent changes in physical and chemical processes within the bed are mediated mainly through changes of the velocity and pressure field brought about by local variations of the void fraction [23] and/or local variations of the particle size. The primary task is therefore to find the relationship between the local resistance of the bed and the local velocity of the flow.

The role of the void fraction in blast furnaces charges and in sinter beds is of special importance for the economy of the process and therefore attempts to describe accurately the conditions within the complex charge are extensive and have appeared over a long period of time, see e.g. papers of Sonntag [146], Mitchell [147], Scheidig and Gruson [148], and Rumpf and Gupte [149].

The purposeful structuralization of the bed has been applied so far only in the blast furnace operation [37], although, as will be seen from this chapter, potential advantages of structuralization of the bed are not by any means confined to metallurgy.

3.1. EQUATIONS GOVERNING THE VELOCITY AND PRESSURE DISTRIBUTION IN INHOMOGENEOUS BEDS

Considerable progress in relating the pressure losses to the velocity in packed beds has been achieved by the work of Ergun [50]. Surveying the work of many of his predecesors Ergun concluded that the factors to be considered are:

The rate of fluid flow, viscosity and density of the fluid, closeness and orientation of packing and size and shape and surface of the particles. These factors thus relate to the fluid as well as the solids.

The principal contribution of Ergun was that he recognized the additive nature of the viscous and kinetic energy losses and using the concept of the mean hydraulic resistance found true dependences of the kinetic and viscous contributions on the void fraction. In his interpretation the conventional friction factor, defined as the ratio of the pressure drop to the energy terms, is a linear function of the velocity and, expressed in terms of the hydraulic mean radius, no longer a function of the void fraction.

The Ergun equation has become a classic correlation and in the extent of the experimental material on which it has been tested has no parallel. These properties have made the Ergun equation attractive for modifications permitting calculation of the velocity and pressure distributions under conditions of non-uniform flow [10,14]. A review of the problem of fluid flow in packed beds has been published by Ziolkowska and Ziolkowski [143].

For the generally non-uniform distribution of resistance in the bed to the flow one has to expect the velocity to be a vectorial quantity. Upon confining the validity of the original scalar form of the Ergun equation to a differential volume of the bed and to the direction of the flow within this volume one can write the following differential and vectorial form of the Ergun equation [16,33,75,149].

$$- \operatorname{grad} P = \mathbf{V} (f_1 + f_2 V) \tag{3-1}$$

The bold set quantities are used to designate vectors. The velocity thus appears in the above equation as the vector $\mathbf{V}$, and its magnitude V. The resistance coefficients f_1, f_2 are identical with the original scalar form of the Ergun equation:

$$f_1 = 150 \, \mu \, (1 - \varepsilon)^2/(d^2 \varepsilon^3) \tag{3-2}$$

$$f_2 = 1.75 \, \varrho \, (1 - \varepsilon)/(d \varepsilon^3) \tag{3-3}$$

but, in contrast, are taken to be functions of the spatial coordinates through the generally spatially variable properties of the matrix of solids and the fluid. Let us note that the particle diameter, in accord with the original Ergun paper, is defined as 6 times the volume of the particle divided by the surface of the particle.

The scalar velocity in the original Ergun interpretation was the velocity of the fluid taken at average pressure. It needs no additional assumption if this velocity is replaced by the vector of the velocity at the pressure and temperature prevailing in the differential volume of the bed.

Of course, the volume considered here must be small to be a differential volume of the bed, yet large enough compared to the size of a single particle.

Under these circumstances Eq.(3-1) represents a balance of forces for this volume. Based on this consideration, Kitaev et. al [66] suggested that an inertial term should be added to the equation as a logical extension by adding inertial terms to the vectorial form of the Darcy equation. The Darcy equation is essentially Eq.(3-1) without the second right-hand side term. In connection with the differential and vectorial form of the Ergun equation, however, the question arises whether the inertial losses had not been already accounted for in the second right hand side term of this equation. Choudhary et al. [79] studied this case numerically and found that the inertial term introduces negligible changes in the computed results, excepting perhaps regions of sudden changes of the resistance.

Another situation where the inertial term could play a role is the flow in packed beds near narrow constrictions. This situation arises, for instance, in beds fed by gas through narrow nozzles. This situation has been analyzed very carefully theoretically and experimentally by Fukutake and Okabe [80] who proposed two forms of the Reynolds number for the bed and the particles in order to preserve similarity between the model and the full scale system. They also found the effect of the bed Reynolds number on gas pressure to be small but definite near the inlet. Eq.(3-1) is thus satisfactory, except perhaps when extremely accurate pressure distribution near the gas inlet is required.

Inspection of Eq.(3-1) shows that the velocity vector should be always parallel to the gradient of pressure. This is, of course, not generally true in unisotropic media. In this case the proportionality factor between the velocity and pressure gradient should have the character of a second order tensor. A generalized form for the linear Darcy form of the flow equation in porous media has been derived by Larson [144].

Yoshizawa [67] tried to construct an additional term to Eq. (3-1) to give the proportionality factor tensorial character on the following assumptions:

1) The tensor has two principal directions given by the velocity vector and the gradient of the hydraulic radius $d\varepsilon/(1 - \varepsilon)$.

2) In the isotropic case, when the gradient of hydraulic radius is zero, the tensor reduces to the scalar in the bracket on the right hand side of Eq.(3-1).

3) If the gradient of hydraulic radius is parallel to the velocity vector, the pressure gradient is also parallel to the velocity.

The term added by Yoshizawa [67] gave Eq.(3-1) the following form

$$- \operatorname{grad} P = \boldsymbol{V} \cdot \left\{ f_1 \boldsymbol{I} + f_2 \boldsymbol{V}\boldsymbol{I} + f_3 \frac{1}{\varepsilon^2} \left(\operatorname{grad} \left(\frac{d\varepsilon}{1 - \varepsilon} \right) \boldsymbol{V}^{\mathrm{T}} + \boldsymbol{V} \left[\operatorname{grad} \left(\frac{d\varepsilon}{1 - \varepsilon} \right) \right]^{\mathrm{T}} \right) \right\} \tag{3-4}$$

where the superscript T designates transpose vectors and $\boldsymbol{I}$ is the identity tensor.

Clearly, if the gradient of the hydraulic resistance equals zero the added term entirely vanishes. If the gradient of the hydraulic radius is parallel to the velocity vector, the above term reduces to a scalar and thus the gradient of pressure and the velocity vector are parallel. Nevertheless, the proportionality factor between these two vectors is then different from that given by Eq.(3-1).

Even Yoshizawa [67] himself found it difficult to evaluate the coefficient f_3 and concluded that the truly tensorial form of the Ergun equation plays a minor role in affecting the pressure losses and the velocity field, except, perhaps in the neighborhood of sharp boundaries between regions of different hydraulic radius. This means between regions of different particle size and/or void fraction.

Inspection of Eq.(3-1) shows that in this form it is inconsistent with the no slip condition, i.e. zero velocity at the wall of the containing surface. As will be seen later the zero velocity at the wall has been observed by several authors performing velocity measurements of the stream of gas issuing from the packed bed.

Brinkman [81] resolved the problem earlier in the case of the Darcy equation by adding Darcy's term to the Stokes equation The appearance of the second--order derivative of velocity removed the problem of inconsistency with the no slip condition and Lebon and Cloot [82] later provided thermodynamic justification and the range of applicability for this form of the Darcy law.

Chandrasekhara and Vortmeyer [83] followed a similar path and added the viscous resistance term to the differential and vectorial form of the Ergun equation to obtain

$$\mu \,\text{divgrad}\, \boldsymbol{V} - \text{grad}\, P = \boldsymbol{V}\,(f_1 + f_2 V) \tag{3-5}$$

They showed the length scale of the viscous resistance term to be $\sqrt{\mu/f_1}$, i.e. small compared to the particle size and confined to a thin boundary layer near the wall where the friction effects may dominate.

It must be noted that the definitions of the coefficients in the Ergun equation, Eqs(3-2), (3-3), incorporate explicit functional dependences on the particle diameter and the void fraction. These functional dependences were constructed on the basis of certain geometrical considerations, namely the randomness of the bed and the concept of the mean hydraulic radius. These assumptions are without doubt seriously violated in the proximity of the wall (see e.g. Chu and Ng [310]) and in general in beds exhibiting small bed to particle diameter ratio. In these cases the functional dependences of the coeficients f_1 , f_2 on particle size and void fraction have to be redefined in accord with a suitable model of the structural inhomogeneity in the proximity of the centaining surface. Such models have been presented in the previous chapter.

An attempt in this direction has been made by Cohen and Metzner [91] who divided the bed into the bulk, transition and the wall region while properly accounting for the void fraction variations, including oscillations, in the expression for the hydraulic radius.

Also, Delmas and Froment [109] extended the Ergun equation in modifications of Mac Donald [110] and Foscolo [111] to account for the tortuosity of the path through the packed bed, the variable cross section of these paths and void fraction oscillations. Their [109] equation expanded by a viscous term, as done earlier by Chandrasekhara and Vortmeyer [83], see Eq.(3-5), in a two dimensional form was then as follows

$$-\frac{\mathrm{d}P}{\mathrm{d}z} = 0.20\left(1 + 3.55\,\frac{1-\varepsilon}{\varepsilon^3}\right)\frac{\varrho V^2}{d} + $$

$$+ 7.53\,\frac{\varepsilon}{1-\varepsilon}\left(1 + 4\,\frac{1-\varepsilon}{\varepsilon^3}\right)\frac{\mu V^2}{d^2} + \frac{1}{r}\frac{\partial}{\partial r}\left(\mu\,\frac{\partial V}{\partial r}\right) \tag{3-6}$$

A disadvantage of this equation is that it does not forsee any radial flow and expects pressure equalization at every horizontal level. Nevertheless, modification of Eq.(3-6) to a vectorial form of the type of Eq.(3-1) should not be difficult.

Solutions of Eq.(3-6) provided oscillatory radial velocity profiles with maxima almost independent of the mean velocity while the peaks increased in height with increasing velocity even in the dimensionless form when scaled by the mean velocity of the flow.

Incorporation of the effect of the containing wall as a contributor to the extent of the solid/liquid surface area and hence the hydraulic radius is, however, considerably more difficult. This is so because the differential and vectorial form of the Ergun equation, in essence, views the bed as a pseudohomogeneous medium and thus cannot answer the question of what happens on an isolated surface.

Similarly as we do not ask what happens on the surface of a single particle of the bed, as it is below the "resolution" of the model equation, the Ergun equation will not provide an answer to the question of what happens on the containing wall. The Chandrasekhara and Vortmeyer equation (3-5) is thus useful particularly to study the phenomena concentrated in the wall region, such as heat transfer, etc.

Mixtures of particles

The above equations are, of course, valid for monosized particles. If the bed consists of particles of different size, difficulties may arise as to how to evaluate the coefficients in Eqs(3-2) and (3-3). Of particular concern is to find suitable values of particle size and void fraction and Shurman [164], for instance, related the permeability of the blast furnace burden to mechanical properties of coke.

Detailed analysis of this problem is beyond the scope of this book. Let it suffice to say that, according to Ergun [107] and Jeschar [84,151], the mean particle size, or its inverse value, may be computed as a mean of the inverse diameters of individual fractions composing the mixture weighted by the fractional volume of the given fraction in the mixture. The void fraction to be used is the void fraction of the mixture. In addition, Jeschar [84] multiplies the resistance coefficient by the ratio of the void fraction of a unisized bed, formed only by the smallest particles in the mixture, to the void fraction of the mixture, this ratio being raised to the power 0.75.

Still the question remains of the void fraction of the mixture, ε_m, as long as it had not been obtained experimentally or from a model as proposed by Dodds [132]. Two component mixtures of particles tend to display minima on the void fraction versus the fraction of the smaller particles in the mixture, X_s. If the fraction of the smaller particles in the mixture is less than that of the mixture exhibiting minimum void fraction, one can visualize the situation as follows: The smaller particles fill the voids of the bed consisting of the larger particles. Asymptotically this segment of the dependence can be expressed by

$$\frac{\varepsilon_m}{\varepsilon} = 1 - \frac{X_s}{1 - X_s} \frac{1 - \varepsilon}{\varepsilon} \tag{3-7}$$

where ε is the void fraction of unisized bed. Naturally, the last equation holds better for X_s near zero and for large particle diameter disparities, when the smaller particles can actually fit into the voids between the larger particles.

Real binary mixtures, however, in this region tend to deviate to higher void fractions as the smaller particles not only fill the pores of the matrix of the larger particles, but also disturb the matrix itself. The disturbance is, of course, greater for smaller ratios of diameters of particles forming the mixture.

In fact, for a certain value of ratio of particle diameters the smaller particles do not fit into the pores of the matrix; this value being sometimes called the "critical occupation ratio". For spheres this critical ratio ranges approximately between 1.36 and 2.41 for the dense and loose packings of the matrix of larger spheres. However, a more stringent condition is imposed by the narrowest size of the pore determining whether the particle can enter the void in the matrix of larger particles by "filtration". This critical value of the ratio of particle diameters forming the mixture is termed the "ratio of entrance". Again for a bed of spheres in a triangular pitch this ratio is 6.46. Uniform binary mixtures of particles in the range of diameter ratios between the above limits would have to be therefore prepared by mixing the two components.

On the other hand, if the fraction of the larger particles in the mixture is small and less than that of the mixture of minimum void fraction, one can visualize the larger particles as being embedded in the bed of smaller particles. The asymptote for this model of the mixture is as follows.

$$\frac{\varepsilon_m}{\varepsilon} = \frac{X_s}{1 + \varepsilon X - \varepsilon} \tag{3-8}$$

Upon equating the last two equations one can obtain the expression for the composition of the mixture of minimum void fraction

$$X_s = \frac{\varepsilon}{1 + \varepsilon} \tag{3-9}$$

while the minimum void fraction amounts to

$$\varepsilon_m = \varepsilon^2 \tag{3-10}$$

In general, however, the void fraction of a binary mixture is a function of both the size ratio and the volume fraction of the components.

Abe et al. [85] found that for technical nonspherical particles such as calcite one can compute the resistance of a mixture on the basis of the model of complete mixing (average void fraction and weighted inverse particle diameter mean) regardless of the particle size ratio, composition of the mixture and packing method. For spherical particles, however, the same authors [85] found that this method may be applied only in region where the volume fraction of the smaller particles exceeds that of the minimum void fraction. Below this limit the state of the mixture depends on the packing method and particle diameter ratio.

For irregular particles, such as e.g. coke, Standish and Borger [124] found that binary mixtures display void fraction minima over a range of compositions, by comparison with spheres where minimum void fraction occurs at a single composition. This is ascribed to the irregularity of the shape of particles hindering normal saturation effects. Analogous behaviour has been reported for ores and sinters.

In ternary mixtures of spheres the minimum void fraction occurs always for the binary mixture of the largest and the smallest spheres. The lines of constant void fraction, drawn in a triangular chart where the corners represent a single component bed, radiate from the of the minimum void fraction toward the medium size sphere corner. Standish and Borger [124] found an analogous behaviour for ternary mixtures of coke. In contrast to spherical particles, however, one must expect higher void fractions in absolute terms than those of comparable mixtures of spheres. This can be explained by the general observation that the void fraction of beds of particles of different shape increases with decreasing particle sphericity. This fact has been best exemplified for a bed of parallelepidal particles of low thickness-to-side ratio such as wood chips in the paper pulp making by Comiti and Renaud [313].

In multicomponent mixtures the minimum void fraction should occur near the binary composition of the largest and the smallest size component of the mixture. A relatively simple expression for the mean void fraction of a packing without macropores for polydisperse particles or discrete size distribution was published by Ouchiyama and Tanaka [155]. The reader can find more details in a handbook by Cumberland and Crawford [129].

3.1.1. Velocity and pressure field of an incompressible fluid in an inhomogeneous bed

The differential and vectorial form of the Ergun equation (3-1) contains two independent variables, namely the velocity vector and pressure. Elimination of pressure can be achieved by applying the vectorial operator rot which yields the following equation

$$\text{rot}\,\boldsymbol{V} - \boldsymbol{V} \times \text{grad}\,[\ln(f_1 + f_2 V)] = 0 \tag{3-11}$$

in which the symbol $\times$ designates the operation of the vector product.

For the generally inhomogeneous distribution of the resistances in the bed the velocity vector is three dimensional and may be written, for instance, in cylindrical coordinates, as

$$\boldsymbol{V} = \mathbf{e}_r V_r + \mathbf{e}_\theta V_\theta + \mathbf{e}_z V_z \tag{3-12}$$

where subscripted $\boldsymbol{e}$'s designate unit vectors in the direction of cylindrical coordinates.

The absolute magnitude of the velocity vector (symbol V) is then given by

$$V = \sqrt{V_r^2 + V_\theta^2 + V_z^2} \tag{3-13}$$

The vector equation (3-11) represents in fact three scalar equations for the three components of the velocity vector. In cylindrical coordinate these three equations take the following form

$$\frac{1}{r}\frac{\partial V_z}{\partial \theta} - \frac{\partial V_\theta}{\partial z} + \frac{1}{f_1 + f_2 V}\left\{\frac{V_z}{r}\left(\frac{\partial f_1}{\partial \theta} + f_2\frac{\partial V}{\partial \theta} + V\frac{\partial f_2}{\partial \theta}\right) - \right.$$

$$\left. - V_\theta\left(\frac{\partial f_1}{\partial z} + f_2\frac{\partial V}{\partial z} + V\frac{\partial f_2}{\partial z}\right)\right\} = 0 \tag{3-14a}$$

$$\frac{\partial V_r}{\partial z} - \frac{\partial V_z}{\partial r} + \frac{1}{f_1 + f_2 V}\left\{V_r\left(\frac{\partial f_1}{\partial z} + f_2\frac{\partial V}{\partial z} + V\frac{\partial f_2}{\partial z}\right) - \right.$$

$$\left. - V_z\left(\frac{\partial f_1}{\partial r} + f_2\frac{\partial V}{\partial r} + V\frac{\partial f_2}{\partial r}\right)\right\} = 0 \tag{3-14b}$$

$$\frac{1}{r}\left(\frac{\partial}{\partial r}(rV_\theta) - \frac{\partial V_r}{\partial \theta}\right) + \frac{1}{f_1 + f_2 V}\left\{V_\theta\left(\frac{\partial f_1}{\partial r} + f_2\frac{\partial V}{\partial r} + V\frac{\partial f_2}{\partial r}\right) - \right.$$

$$\left. - \frac{V_r}{r}\left(\frac{\partial f_1}{\partial \theta} + f_2\frac{\partial V}{\partial \theta} + V\frac{\partial f_2}{\partial \theta}\right)\right\} = 0 \qquad \text{(3-14c)}$$

Upon confining, for the time being, the considerations to incompressible fluids, the velocity vector, i.e. its components must satisfy the continuity equation. For an incompressible fluid this equation takes the form

$$\text{div } \boldsymbol{V} = 0 \qquad \text{(3-15)}$$

Upon obtaining solution of the velocity field from Eq(3-11) one should, in principle, obtain the pressure field from the original equation (3-1), utilizing now the already known field of the velocity vector $\boldsymbol{V}$. It is, however, more advantageous to solve the pressure field from the following equation obtained after applying the operator div on Eq.(3-1) and making use of the continuity equation (3-15) with the result

$$\text{divgrad } P = -\boldsymbol{V} \cdot \text{grad } (f_1 + f_2 V) \qquad \text{(3-16)}$$

The dot in this equation designates the operation of the scalar product.

On the contrary, one can, of course, follow an alternative procedure by eliminating the velocity from the original equation (3-1) using vectorial algebraic operations to obtain the following partial differential second order equation

$$\text{divgrad } P - \text{grad } P \cdot \text{grad}\left[\ln\left(\frac{f_1}{2} + \sqrt{\left(\frac{f_2}{2}\right)^2 + f_2 \text{ grad } P}\right)\right] = 0 \quad \text{(3-17)}$$

The quantity grad P under the under the square root sign represents again the magnitude of the vector grad P.

Solution of the problem via Eq. (3-17) is, however, less advantageous. Firstly because the information about the velocity field appears primary from the stand point of interactions between the structure of the bed and the thermal and chemical process taking place within the bed. Secondly, Eq. (3-17) appears to have, in comparison with Eq. (3-16), less favourable properties from the standpoint of convergence of the numerical method of solution. The observed differences in the number of necessary iterations are of one order of magnitude [16].

In a number of cases the inhomogeneity of the structure of the bed displays a certain symmetry. The functions f_1, f_2 then display the same symmetry properties, which in turn reflect in the symmetry of the field of the velocity.

A practical example of such a symmetry may be the increased void fraction near the containing wall of the column (reactor, bed, furnace, etc.), which had been theoretically analyzed in previous paragraphs. In the given case of a cylindrical container we are dealing with a cylindrical symmetry which will manifest itself in the velocity field having only two components nonvanishing.

The characteristic nonzero components in case of axial symmetry are the axial velocity, V_z, and the radial velocity component, V_r. These two components are constrained by the equation of continuity and the problem may be transformed into one of solving a single scalar second-order partial differential equation. The order of this equation is thus higher than the order of the original equation.

Upon introducing a new quantity, the stream function, ψ, in such a way that in each point of the plane of the coordinates r, z, the velocity vector is tangential to the contour of constant stream function, the relationship between the stream function and the two nonvanishing velocity components is given by the following expressions

$$V_z = \frac{1}{r}\frac{\partial \psi}{\partial r} \tag{3-18}$$

$$V_r = -\frac{1}{r}\frac{\partial \psi}{\partial z} \tag{3-19}$$

where r and z are corresponding cylindrical coordinates.

The advantage of using the stream function is that the equation of continuity is automatically satisfied. If the axially symmetric flow through the nonuniform bed satisfies further the following inequality

$$V_0 f_2/f_1 \doteq 0.07\ Re \gg 1 \tag{3-20}$$

the differential equation for the stream function, obtained by substituting Eqs (3-18), (3-19) into (3-11), simplifies to the two-dimensional form with the following result

$$\frac{\partial^2 \psi}{\partial r^2}\left[2\left(\frac{\partial \psi}{\partial r}\right)^2 + \left(\frac{\partial \psi}{\partial z}\right)^2\right] + \frac{\partial^2 \psi}{\partial z^2}\left[2\left(\frac{\partial \psi}{\partial z}\right)^2 + \left(\frac{\partial \psi}{\partial r}\right)^2\right] +$$

$$+ \left[\left(\frac{\partial \psi}{\partial r}\right)^2 + \left(\frac{\partial \psi}{\partial z}\right)^2\right]\left[\frac{\partial \psi}{\partial z}\frac{\partial \ln f_2}{\partial z} + \frac{\partial \psi}{\partial r}\left(\frac{\partial \ln f_2}{\partial r} - \frac{2}{r}\right)\right] +$$

$$+ 2\frac{\partial^2 \psi}{\partial r \partial z}\frac{\partial \psi}{\partial r}\frac{\partial \psi}{\partial r} = 0 \tag{3-21}$$

The inequality (3-20) restricts the validity of Eq.(3-21) to the case of asymptotically fast flows with prevailing inertia energy dissipation. The Reynolds number in the inequality (3-20) is defined on the basis of the mean velocity V_0 and the inverse of the specific surface of particles as the length scale. For *Re* exceeding 150 the contribution of the viscous term in the Ergun equation will be less than 10 %.

Also one can similarly obtain corresponding form of Eq. (3-11) in terms of the stream function for the asymptotically slow flows with prevailing viscous dissipation. Analogously for Re less than 1.5 the inertia contribution to the overall dissipation is less than 10 %.

The advantage of both asymptotic forms of the flow equation is that the solution in terms of the dimensionless velocity vector, expressed as a fraction of the mean velocity V_0, is invariant to V_0. This fact considerably increases the generality of the conclusions obtained on the basis of these asymptotic solutions.

The complete form of Eq.(3-11), transformed by means of the stream function, is then necessary only in the transition region between the two above indicated limits of the Reynolds number ($1.5 < Re < 150$).

Certain precautions are necessary in expressing the vector operators in curvilinear coördinates and also for other types of symmetry. The rules for the simple operations between vectors are not generally valid between these operators.

In order that we may solve Eq.(3-21), or its equivalents for the viscous or transition region, one has to still specify the boundary conditions. With the concept of the stream function these conditions may be specified as follows:

The confining surfaces of the bed, such as e.g. the walls, are impervious to the flowing fluid and the velocity vector is therefore parallel to these surfaces. Corresponding mathematical formulation of this fact is that the stream function is constant at the containing surfaces. The same is true about the axis of the bed if the problem in question is axially symmetric. The axis is also impervious to the flow and the stream function also remains constant in this case.

A macroscopic expression for the law of conservation of mass (or volume in case of an incompressible fluid) is the same flow rate of the fluid between the containing surfaces, or between the walls and the axis, in case of axially symmetric flows, at any horizontal level. This flow rate then determines the difference between the values of the stream function on the wall and the axis. One of these values of the stream function may be assigned an arbitrary value (preferably zero) as we have increased the order of the governing differential equation. The other extreme value of the stream function is determined by integration of, for the case of axial symmetry, Eq. (3-18) between the two extreme values of radius.

The surfaces through which the fluid enters and leaves the bed, i.e. mostly horizontal planes of the top of the packed layer and the supporting grid, are isobaric surfaces as the space above and below the bed proper has a negligible resistance compared to the bed itself. With respect to isobaric surfaces the gradient of pressure vector, grad P, is perpendicular and from the basic equation (3-1) then follows that also the velocity vector is perpendicular to this surface. In cylindrical coordinates we then have that only the axial velocity component, V_z, is nonzero in this case.

This conclusion is valid even if the flow exhibits no axial symmetry and there is a nonzero tangential velocity component within the bed. For axially symmetric flows one can then derive from the condition of zero radial velocity component, V_r, on the isobaric outlet and inlet surfaces using Eq. (3-19) that the derivative of the stream function in the direction perpendicular to the isobaric surface is zero. Thus $\partial\psi/\partial z = 0$.

3.1.2. Energy dissipation in an incompressible fluid flowing through an inhomogeneous bed

An expression for the rate of energy dissipation in the fluid flowing through the bed is obtained [16] as a scalar product of the differential vectorial form of the Ergun equation (3-1) and the velocity vector with the following result:

$$\mathbf{V} \cdot \text{grad } P = -(f_1 V^2 + f_2 V^3) \tag{3-22}$$

Since, however, we may write that:

$$\text{div } (\mathbf{V}P) = \mathbf{V} \cdot \text{grad } P + P \text{ div } \mathbf{V} \tag{3-23}$$

and for an incompressible fluid the second term on the right hand side of Eq.(3-23) vanishes the expression for the rate of energy dissipation (3-22) may be modified finally to:

$$\text{div } (\mathbf{V}P) = -(f_1 V^2 + f_2 V^3) \tag{3-24}$$

Upon integrating the last equation over the whole volume of the layer, O, while designating its total external surface by S, we obtain the following:

$$\iint_S P \mathbf{V} \cdot \text{d}\mathbf{S} = -\iiint_O (f_1 V^2 + f_2 V^3) \, \text{d}O \tag{3-25}$$

in which we applied the Gauss-Ostrogradskii theorem for the transformation of the volume integral over the volume of the layer to the surface integral over the whole external surface of the bed (container).

On the left hand side of Eq.(3-25) the element of the surface of the bed, d$\mathbf{S}$ is taken as a vector whose magnitude equals the surface area of the element and its direction is given by the outer normal to the surface.

In the previously examined case of a cylindrical container (column, reactor) a nonzero contribution to the surface integral on the left hand side of Eq.(3-25) provide only the grid on which the layer rests and the top surface of the layer. On the jacket of the cylinder, as follows from the boundary conditions, the velocity vector is parallel to the surface confining the layer and hence the scalar product behind the integration sign on the left hand side of Eq.(3-25) is here zero.

Expressed by words, the integral on the left hand side represents the rate of change (with respect to the orientation of the vector $d\mathbf{S}$ decrease) of pressure energy in a unit volume of the bed. The integral on the right hand side of the same equation expresses the rate of energy dissipation in the whole volume of the bed.

3.1.3. Parallel flow and the minimum of energy dissipation flow distribution

Parallel flow with axial symmetry

Let us examine now the case of the flow through an inhomogeneous layer under certain conditions permitting us to circumvent numerical methods of solution of Eq.(3-11) or Eqs. (3-14a,b,c). Consider the case of the flow in a cylindrical container when the resistance to the flow varies radially, yet the radial profiles of the resistance are on all horizontal levels the same. This means that the derivative of the resistance with respect to the axial coordinate vanishes in the whole volume of the bed.

The special case of such a layer is shown in Fig. 3-1 depicting a cylindrical container A with a layer of such properties that within a hypothetical cylinder B, concentric with the container, the layer has generally different resistance from that in the region outside the hypothetical cylinder.

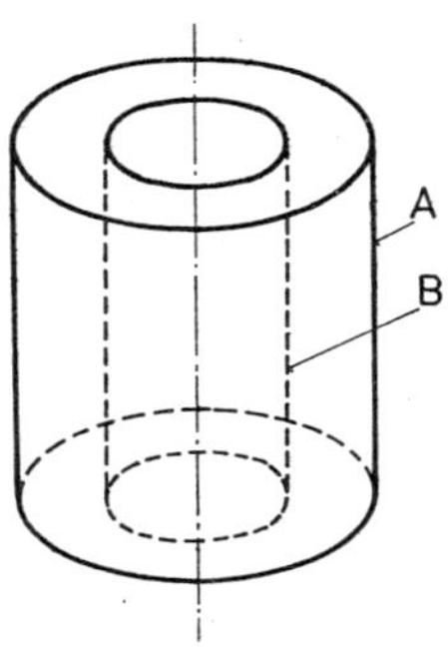

Fig.3-1
Sketch of the layer with a hypothetical cylindrical surface separating regions of different resistance. A - container wall; B - hypothetical concentric cylindrical surface.

From the practical standpoint this may mean, for instance, that the peripheral region outside the hypothetical cylinder B is the region of increased void fraction due to the structural defects of the bed near the walls, while the core is the region with the void fraction corresponding to a homogeneous random bed.

Additional considerations may be extended to the case of an arbitrary axially symmetric radial profile of the resistance. If, as before, this resistance does not change axially the flow through the layer, under isobaric conditions at the inlet and outlet, will be nonuniform but parallel. This means that the velocity vector, $\boldsymbol{V}$, has a single nonvanishing component in the direction of the axis z, i.e. V_z and this component is a function of the radial coordinate, r, only, i.e. $V_z = V_z(r)$. Eqs (3-14a,b,c) under these circumstances reduce to a single equation of the form:

$$\frac{\partial V_z}{\partial r} + V_z \frac{\partial(\ln(f_1 + f_2 V)}{\partial r} = 0 \tag{3-26}$$

Since V_z is the only nonzero component of the velocity vector, its absolute value, V, appearing in the second term of Eq. (3-26) equals this component, i.e. $V = V_z$. After this we can integrate Eq (3-26) with the result:

$$V_z = -\frac{f_1}{2f_2} + \sqrt{\left(\frac{f_1}{2f_2}\right)^2 + \frac{k_1}{f_2}} \tag{3-27}$$

where k_1 is an integration constant. Its value may be determined from the material balance on the fluid by substituting Eq.(3-27) into the following equation:

$$V_0 R^2 = \int_0^R 2rV_z \, \mathrm{d}z \tag{3-28}$$

where R designates the radius of the bed.

From Eq.(3-25) it is apparent that the rate of energy dissipation per unit height of the bed, E, may be in this case expressed as:

$$E = \int_0^R (f_1 V_z^2 + f_2 V_z^3)\, 2\pi r \mathrm{d}r \tag{3-29}$$

Now we may ask what is the radial distribution of velocity, $V_z = V_z(r)$, corresponding to the given radial profile of the resistances f_1, f_2, under which the rate of energy dissipation is minimal? Thus we seek an extreme of the integral (3-29) while the value of the integral on the right hand side of Eq.(3-28) must remain constant. A necessary condition for such an extreme is that the first variation of the linear combination of both mentioned integrals be equal to zero.

The first variation of the integral

$$\int_0^R (f_1 V_z^2 + f_2 V_z^3 - k_2 V_z)\, 2\pi r dr \qquad (3\text{-}30)$$

where k_2 is an arbitrary constant, equals zero if we have

$$3f_2 V_z^2 + 2f_1 V_z - k_2 = 0 \qquad (3\text{-}31)$$

which occurs when

$$V_z = -\frac{f_1}{3f_2} + \sqrt{\left(\frac{f_1}{3f_2}\right)^2 + \frac{k_2}{3f_2}} \qquad (3\text{-}32)$$

The constant k_2 could be also determined from Eq.(3-28) but a comparison of Eq.(3-31) with (3-27) shows that they are not identical. The distribution of velocities under parallel flow through an inhomogeneous layer is thus not identical with that corresponding to the minimum rate of energy dissipation in the layer.

The relative contribution of both terms on the right hand side of Eq.(3-1) to the energy dissipation in the bed is given approximately by the following ratio

$$V_0 f_2/f_1 \doteq Re/15 \qquad (3\text{-}33)$$

This ratio earlier yielded the inequality (3-20) and the limiting values of the Reynolds number (150 or 1.5) making always one of the contributions at least an order of magnitude larger. In the transition region between both limiting values of the Reynolds number both contributions are significant and, due to the differences between Eqs(3-27) and (3-32), the rate of energy dissipation under the parallel flow is higher.

For fast flows, with the dissipation predominantly due to the inertia effects, there follows from Eq.(3-27), after expression of the constant k_1 from Eq. (3-28), the following distribution of velocities under the parallel flow

$$V_z = V_0 f_2^{-0.5}/(f_2^{-0.5})_0 \qquad (3\text{-}34)$$

where

$$(f_2^{-0.5})_0 = \frac{2}{R^2}\int_0^R \frac{r\, dr}{f_2^{-0.5}} \qquad (3\text{-}35)$$

For asymptotically slow flows with prevailing viscous dissipation, there follows from the solution of Eq.(3-26) after putting $f = 0$, the following velocity distribution

$$V_z = V_0 f_1^{-1}/(f_1^{-1})_0 \qquad (3\text{-}36)$$

where

$$(f_1^{-1})_0 = \frac{2}{R^2}\int_0^R \frac{r\,\mathrm{d}r}{f_1} \tag{3-37}$$

One can easily show that after determination of the constant k_2 in Eq.(3-32) using Eq.(3-28), there result for both asymptotic cases expressions identical with those in Eqs(3-34) and (3-36). In both limiting cases thus the distribution of velocities of the fluid in a bed with an inhomogeneity leading to the parallel flow is dentical with the distribution providing for the minimum rate of energy dissipation.

This finding appears more natural even though the minimum rate of energy dissipation by itself is not a physical law. Higher than minimal rates of energy dissipation, predicted by Eq. (3-27) in the transition region, probably therefore reflect the empirical character of the Ergun equation.

From the asymptotic expressions for the distribution of velocities in an inhomogeneous layer under the conditions of the parallel flow it follows that the same relative inhomogeneities of the profile of resistance will cause stronger nonuniformity of the velocity profile at low flow velocities when viscous dissipation prevails.

Parallel flow under angular distribution of inhomogeneity

Parallel flow, however, may occur also in beds exhibiting other properties as far as the distribution of the local resistance is concerned, so that that axial velocity component is a function only of the angular coordinate. A simple case of this type is shown in Fig.3-2 where a cylindrical bed has a uniform resistance throughout except for a segment within which the resistance is different but also constant.

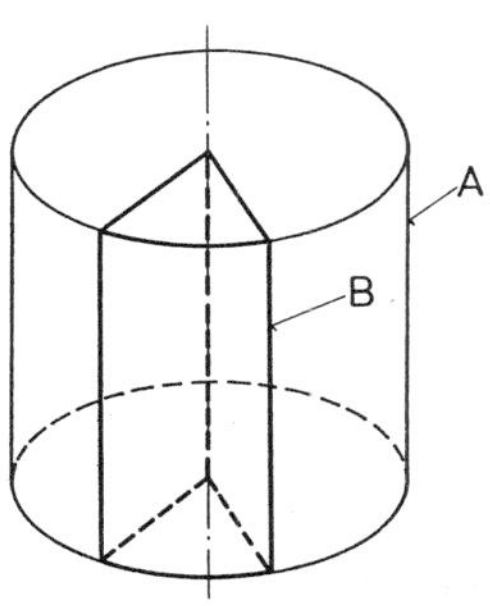

Fig.3-2
Sketch of the layer with hypothetical planes separating segment of the bed of different resistance. A - container wall; B - hypothetical segment.

For this type of structural nonuniformity, given again by different particle size and/or void fraction, we can write [35]

$$\frac{\partial f_1}{\partial z} = \frac{\partial f_2}{\partial z} = \frac{\partial f_1}{\partial r} = \frac{\partial f_2}{\partial r} = 0 \quad \text{for all} \quad z \tag{3-38}$$

This means that the resistance varies only with the angular coordinate and this distribution is the same at all horizontal levels.

If the inlet cross section is an isobaric surface we find that the only non-zero velocity component is V_z which varies only with the angle θ. Substituting these finding into Eq.(3-14a) we get

$$\frac{\partial \ln V_z}{\partial \theta} + \frac{\partial \ln (f_1 + f_2 V_z)}{\partial \theta} = 0 \tag{3-39}$$

Analogously to Eq.(3-26) we obtain the solution in formally the same form given in Eq.(3-27), where the constant, say k_3, instead of k_1, is given by the balance on the fluid from

$$2\pi V_0 = \int_0^{2\pi} V_z \, d\theta = \int_0^{2\pi} \left(\frac{f_1}{2f_2} + \sqrt{\left(\frac{f_1}{f_2}\right)^2 - \frac{k_3}{f_2}} \right) d\theta \tag{3-40}$$

and cannot be determined unless the resistances f_1, f_2 are known explicit functions of the angle θ. Explicit expressions for the velocity component V_z as a function of the angle θ can be found similarly as in the case of beds with axial symmetry for asymptotic regions of the velocity, i.e. when f_2 is either substantially smaller or, on the contrary, larger than f_2.

Naturally, these equations hold also for more complicated structures of the inhomogeneity exhibiting the same symmetry. Conclusions regarding the minimum rate of energy dissipation with respect to velocity distribution are the same as in the case of axially symmetric flows.

3.1.4. Parallel flow with axial symmetry – experimental results

An example of the parallel flow of practical importance is the flow through a packed cylindrical bed under the presence of the wall effect.

As has been shown the presence of the containing walls induces typical oscillations of the void fraction. Apart from the oscillations the void fraction in the wall region exhibits in general values higher than in the bulk of the bed.

Axial variations of the resistance in a random bed should be negligible and the flow thus should be parallel except perhaps near the inlet and outlet surface. This situation has been studied extensively by measuring the velocity of the gas at the outlet from the bed, particularly its radial profile, and has been correlated in various ways [52,86-94,100,103,106]. In some cases [87,89] the correlations were determined on the basis of the experimental data of Schwartz and Smith [86]. Other authors [90,91,93,100,103,133] did not correlate their data.

Velocity profiles directly within the pores were measured by Van der Merwe and Gauvin [105] in a cubic packing of equal spheres 7 cm in diameter. Earlier, Mickley et.al. [108], using the same technique concluded that the flow leaving the voids in the last row of particles is basically different from the flow inside the packing.

An example of the velocity profile measured by Drahoš et.al. [52], together with the radial profile of the void fraction is shown in Fig.3-3. Based on these results these authors proposed a simple model that divides the bed into two zones. The central zone, or the core, begins 2.25 particle diameter away from the wall and the mean velocity in this core is given by

$$\frac{V_z(\text{core, mean})}{V_0} = 0.0022\,\frac{2R}{d} + 0.83 \tag{3-41}$$

while in the wall zone the velocity has a maximum at the distance 0.25 particle diameters from the wall with the maximum velocity being

$$\frac{V_z(\text{wall zone, max})}{V_z} = 0.018\,\frac{2R}{d} + 1.2 \tag{3-42}$$

The validity of both equations has been limited by the authors [52] to $5 < (2R/d) < 75$.

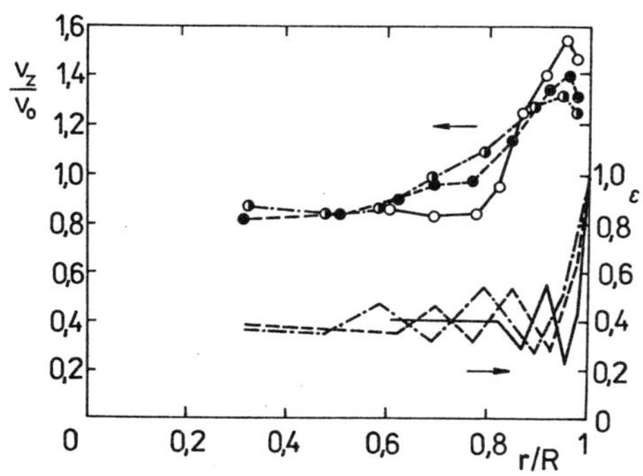

Fig.3-3
Radial profile of the dimensionless axial velocity component obtained by Drahoš and Čermák [52] and corresponding void fraction profile [15] for particle of different size. Column to particle diameter: Empty circles - 22.9; full circles - 13.6; half-empty circles - 9.6.

A simplified zone model, dividing the bed between the wall region and the bulk, proposed by Martin [88] has been described in chapter 2. The same author then calculated the velocities in the two respective regions from the Ergun equation. The ratio of the superficial velocities in the wall and the bulk region is then given by a quadratic equation with the coeficients determined by the coefficients of the Ergun equation (3-2), (3-3). For fast and slow flow, however, this ratio is determined only by the void fractions in the wall (ε_w) and the bulk region (ε_b) respectively

$$\frac{V_z(\text{wall})}{V_z(\text{bulk})} = \left(\frac{1-\varepsilon_b}{1-\varepsilon_w}\right)^2 \left(\frac{\varepsilon_w}{\varepsilon_b}\right)^3 \quad \text{for} \quad Re \to 0 \tag{3-43}$$

$$\frac{V_z(\text{wall})}{V_z(\text{bulk})} = \sqrt{\left(\frac{1-\varepsilon_b}{1-\varepsilon_w}\right) \left(\frac{\varepsilon_w}{\varepsilon_b}\right)^3} \quad \text{for} \quad Re \to \infty \tag{3-44}$$

If the wall region is identified as in Martin's model its void fraction is given by Eq.(2-48) while the void fraction of the bulk is that of the undisturbed bed. The last two equations again demonstrate clearly that the degree of flow nonuniformity increases with decreasing velocity of flow.

Vortmeyer and Schuster [92] solved the Ergun equation, expanded by the viscous resistance term (3-5), in the one dimensional form to obtain the radial velocity profiles for the parallel flow. The void fraction near the wall varied according to Eq.(2-50).

The results were processed into the following correlation

$$\frac{V_z}{V_0} = \beta \left\{ 1 - \exp\left[a \frac{R}{d}\left(1 - \frac{r}{R}\right)\right] \left[1 - n \frac{R}{d}\left(1 - \frac{r}{R}\right)\right]\right\} \tag{3-45a}$$

where

$$\beta = \frac{1}{2}\left(\frac{R}{d}\right)^2 \left[\frac{1}{2}\left(\frac{R}{d}\right)^2 - \frac{\left(n\frac{R}{d} - 1\right)\left(a\frac{R}{d} + 1\right)}{a^2} + \right.$$

$$\left. + n \left(\frac{\left(\frac{R}{d}\right)^2}{a} + 2\frac{\left(\frac{R}{d}\right)}{a^2} + \frac{2}{a^3} \right) - \frac{\exp\left[a\frac{R}{d}\right]}{a^2} \left(1 - n\frac{R}{d} + \frac{2n}{a}\right) \right]^{-1} \tag{3-45b}$$

where

$$a = \frac{4n}{4 - n} \tag{3-45c}$$

and

$$n = 112.5 - 26.31\ Re_p + 10.97\ Re_p^2 - 0.1804\ Re_p^3$$

$$\text{for} \quad 0.1 \leqq Re_p < 1 \tag{3-45d}$$

$$n = -\ 1\,803 + 201.62\ (\ln Re_p + 4) - 3\,737\ (\ln Re_p + 4)^{0.5} +$$

$$+\ 5\,399\ (\ln Re_p + 4)^{1/3} \quad \text{for} \quad 1 \leqq Re_p < 1\,000 \tag{3-45e}$$

$$n = 27 \quad \text{for} \quad Re_p \leqq 1\,000 \tag{3-45f}$$

where the Reynolds number has been defined on the basis of the particle diameter and the mean velocity of the flow.

The velocity profiles based on this equation, of course, do not show any oscillations as the void fraction profile used was also smoothed. They show, however, maxima resulting from the friction on the wall through the presence of the viscous losses. The distance of this maximum is less than one particle radius while the height of the maximum diminishes with increasing Reynolds number.

A relatively simple expression for the velocity profile has been proposed by Fahien and Stankovic [89]. Their equation though does not provide for the change of the velocity profile with the Reynolds number.

Chu and Ng [310] investigated the flow in a packed tube of very small tube to particle diameter ratio on a computer simulated detailed description of its morphology. Their results showed that the presence of the walls has two counteracting effects. Higher porosity promotes higher flow but higher surface area per unit volume near the wall hinders it. For $(2R/d)$ larger than 25 the permeability is the same as that of a large diameter tube. For $(2R/d)$ less than 8 the confining walls will cause a marked increase of the void fraction near the wall and the permeability of the tube is then always larger than that of an unconfined bed. For $(2R/d)$ between 8 and 25 it can be larger or lower than that of an unconfined bed.

In spite of the large amount of experimental work carried out, the question of the relevance of the outlet velocity profile does not seem to be satisfactorily resolved. Experimental methods employing velocity probes just above the bed sense the velocity just above the region, which by itself is not quite representative of the main body of the bed (see the void fraction oscillation profiles near the top of the bed in Fig.(2-4)). In addition, Kirillov et al.[100] found a significant increase of the velocity in the center if the surface of the bed was concave. If the surface was convex there was increased velocity near the wall. No velocity maxima were observed [100] for flat surface of the bed.

The velocity probes usually have to be located at some distance above the bed in order to smooth out individual velocity components. Incidentally, Drahoš et al. [52] and Ziolkowska et al. [96,97,99] were the only ones to measure individual velocity components above the bed. This experimental smoothing procedure thus diminishes the velocity peak near the wall and ultimately wipes out the velocity oscillations which should result from the void fraction oscillations. In fact, Schuster and Vortmeyer [101] showed, by solving the one dimensional form of Eq.(3-5) for the bed and the space just above the bed, the rapid decay of the velocity maximum near the wall. The effect of the distance of the probe was shown experimentally e.g. by Lyckowski [106]. Roughly it can be said that if the probe is too close to the outlet surface the profile reflects more the interstitial velocity distribution. If the probe is too far away, an empty tube velocity profile results. In the former case usually the integral mean value of the obtained velocities does not equal the volumetric flow rate as the non-axial velocity components are still important.

The interstitial velocities in the bulk of the bed and the wall region were found to be statistically the same in a recent paper by Caycik and Gunn [104]. These authors employed a tracer technique in a bed passed by liquid. With this conclusion, of course, the superficial velocities within the bed should reflect the radial void fraction oscillations. The technique employed [104] no doubt reflects the conditions within the bed.

An axially symmetrical parallel flow type bed has been investigated by Choudhary et al [79] and Szekely et al [95]. They studied the flow distribution in a bed composed in the peripheral annular zone (Fig.3-1) of particles three times the diameter of the particles composing the core. They found good agreement between the experimental and computed velocity profiles.

A parallel type of flow in an annular concentric bed has been investigated computationally by Cheng and Hsu [114]. The void fraction in the proximity of both the inner and the outer containing cylindrical surface was described by the Vortmeyer and Schuster [92] exponential void fraction profile, Eq.(2-50). (Here we note that Marivoet et al [133] found oscillating velocity profiles not only near the wall but also near a rod located in the axis of the bed.) The govering equation used [114] was essentially that of Chandrasekhara and Vortmeyer Eq. (3-5) without the inertia term and in the unidimensional form. The velocity distribution profiles obtained depended, of course, on the parameters of the void fraction profiles. Keeping these constant the profiles changed with the particle to inner bed radius ratio and the radius of the outer cylinder. The profiles displayed velocity maxima near the containing surfaces with the maximum near the outer cylinder being slightly higher. The maxima near the outer cylinder were also somewhat farther away from the wall than those near the inner cylinder.

3.2. VELOCITY AND PRESSURE FIELD UNDER THE FLOW OF AN IDEAL COMPRESSIBLE GAS IN AN INHOMOGENEOUS MEDIUM

Since already the principal equation (3-1) was formulated in the differential form, transition to the case of a compressible fluid requires only that we respect corresponding pressure dependences of its physical properties.

From the physical properties appearing in the resistance coefficients f_1, f_2 in Eqs.(3-2) and (3-3) a significant pressure dependence exhibits only density if the flowing fluid is gas. The pressure dependences of viscosity for the pressure gradients occurring in beds of particles, are mostly negligible.

In case of compressible fluids it is convenient to work, in view of the equation of continuity, with mass velocities. The principal equation (3-1) is therefore multipled by density, and, if the flowing fluid is an ideal gas, its behaviour may be described by the following equation

$$\varrho = \varrho_0 P/P_0 \tag{3-46}$$

where ϱ_0 designates density at pressure P_0. With the aid of this equation of state the vectorial equation of motion of a compressible ideal gas may be described as follows

$$\text{grad}\,(P^2) = -\mathbf{G}(b_1 + b_2 G) \tag{3-47}$$

This approach to treating losses of compressible gases has been used already by Ergun [107] in his scalar form of Eq.(3-1) applied to pressure drop in blast furnaces. The bold set symbols are again vectors and the symbol $\mathbf{G}$ designates the vector of superficial mass velocity, i.e. velocity related to empty column cross section. The coefficients b_1, b_2 are defined in the following expression in which the ratio P_0/ϱ_0 is independent of pressure

$$b_1 = 300\mu \frac{P_0}{\varrho_0} \frac{(1-\varepsilon)^2}{d^2\varepsilon^3} \tag{3-48}$$

$$b_2 = 3.5 \frac{P_0}{\varrho_0} \frac{(1-\varepsilon)}{d\varepsilon^3} \tag{3-49}$$

Elimination of the squared pressure term, as one of the dependent variables, is again achieved by applying the vectorial operator rot with the following result

$$\text{rot}\,\mathbf{G} - \mathbf{G} \times \text{grad}\,[\ln(b_1 + b_2 G)] = 0 \tag{3-50}$$

From the known field of interstitial velocity, obtained from the solution of Eq.(3-50), we change to the solution of the squared pressure term from the equation obtained by applying the operator div to Eq.(3-47)

$$\text{divgrad}\,(P^2) = -\mathbf{G} \cdot \text{grad}\,(b_1 + b_2 G) - (b_1 + b_2 G)\,\text{div}\,\mathbf{G} \tag{3-51}$$

This equation is linear with respect to the pressure squared term and more advantageous from the standpoint of the numerical solution than Eq.(3-47).

The equation of continuity has the following form

$$\text{div}\,\mathbf{G} = -\frac{\partial \varrho}{\partial t} \tag{3-52}$$

and hence for steady flows of a compressible ideal gas Eq.(3-51) may be simplified to

$$\text{divgrad}\,(P^2) = -\mathbf{G} \cdot \text{grad}\,(b_1 + b_2 G) \tag{3-53}$$

Scalar product of the vector of superficial mass velocity with Eq.(3-47) yields after some arrangement

$$\text{div}\,(\mathbf{G}P) = -\frac{1}{2P}(b_1 G^2 + b_2 G^3) + \frac{1}{2} P\,\text{div}\mathbf{G} \tag{3-54}$$

In case of the steady flow, again with the aid of the corresponding form of the equation of continuity (3-52) and additional algebra, one can obtain the form with the volume superficial velocity

$$\text{div}\,(\mathbf{V}P) = -\frac{1}{2}\frac{\varrho}{P}(b_1 V^2 + b_2 V^3) + P\,\text{div}\,\mathbf{V} \tag{3-55}$$

In accord with the equation of state (3-46) one may replace the ratio of the density to pressure on the right hand side by the ratio of their reference values and, as the next step, change over to the original coefficients f_1, f_2, defined by Eqs.(3-2) and (3-3). Thus

$$\text{div}\,(\mathbf{V}P) = -(f_1 V^2 + f_2 V^3) + P\,\text{div}\,\mathbf{V} \tag{3-56}$$

Integration of the last equation over the volume of the bed and on transition on the left hand side using the Gauss Ostrogradskii theorem, we find that for an isothermal flow of a compressible ideal gas the rate of conversion of pressure energy in the system equals the rate of energy dissipation in the layer, decreased by the rate of reversible conversion to internal energy of the system. Under the isothermal steady flow, however, the product of the volume velocity and pressure is directly proportional to the mass velocity and its divergence equals zero. The left hand side of Eqs.(3-55) and (3-56) then vanishes and the dissipation takes place exclusively at the expense of internal energy of the flowing gas.

Quite analogously we can extend also the other conclusions regarding the flow of a compressible ideal gas if we replace the coefficients f_1, f_2 by b_1, b_2. We then obtain analogous results, for instance, for the parallel flow as in Eq.(3-27).

3.3. THERMALLY INDUCED GAS FLOW INHOMOGENEITIES IN A FIXED BED

The coefficients of resistance of the differential vectorial form of the Ergun equation (3-1) and (3-47) contain physical properties of the fluid while the arrangement into the form of Eq.(3-47) is based on the properties of an ideal gas.

However, the motion of the fluids through the layer of particles takes place in a number of practical situations under nonisothermal conditions. A typical example is a heat regenerator utilizing the large capacity of the solid particles in comparison with the flowing gas.

Already on the basis of preceding analyses of structural effects on the homogeneity of the flows it is apparent that the nonisothermality of the bed alone, even in a structurally homogeneous layer, will cause deviations from the uniformity of the flow of the fluid. In the following we shall formulate the question of the velocity and pressure field in a nonsiothermal layer of particles. For these purposes we have to define first an explicit dependence of the physical properties on temperature.

If we analyze again the flow of an ideal gas, its state behaviour, written before in Eq.(3-46), may be further modified into en explicit dependence on temperature in the following manner

$$\frac{P}{\varrho} = \frac{P_0}{\varrho_0} T \frac{R_g}{M_g} \tag{3-57}$$

Where T is the absolute (Kelvin) temperature and R_g and M_g designate the versatile gas constant and the molecular weight of the flowing gas.

Also the viscosity of gas, appearing in the resistance coefficients, depends strongly on temperature. From the kinetic theory of gases and using the concept of the gas as being composed of solid spheres with no force interactions, one can derive an expression for the viscosity predicting its increase with the square root of absolute temperature. Collision diameters of the molecule and its mass appear in the coefficient of proportionality.

Experimentally determined values, however, show that the viscosity of gases under low densities changes rather with the 0.6-th up to the first power of absolute temperature. For air around zero degrees Celsius the value of the exponent approaches more to the upper of the indicated limits. At temperature around hundred degrees Celsius the exponent still amounts to about 0.68 and around thousand degrees Celsius the exponent is still about 0.62.

A substantially better prediction of the temperature dependence of viscosity of gases can be achieved on the basis of the Chapman-Enskog theory of monoatomic gases and the potential energy of molecular interactions, that can be determined on the basis of the empirical Lennard-Jones (6-12) potential. For our purposes, however, this approach, in spite of its accuracy, is not convenient as it does not provide an explicit dependence of viscosity on temperature.

We shall therefore start from the following empirical power law temperature dependence of viscosity of gas

$$\frac{\mu}{\mu_0} = \left(\frac{T}{T_0}\right)^a \tag{3-58}$$

where the exponent a assumes some of the above indicated values characteristic for the gas, independently of temperature.

The differential vectorial equation of motion may then be written as follows

$$\text{grad}\,(P)^2 = -\mathbf{G}\left[c_1\left(\frac{T}{T_0}\right)^{a+1} + c_2\frac{T}{T_0}G\right] \tag{3-59}$$

where the coefficients c_1, c_2 are defined as

$$c_1 = \frac{300(1-\varepsilon)^2}{d^2\varepsilon^3}\frac{\mu_0 R_g T_0}{M_g} \tag{3-60}$$

$$c_2 = \frac{3.5(1-\varepsilon)}{d\varepsilon^3}\frac{R_g T_0}{M_g} \tag{3-61}$$

Let us introduce still the ratio of these coefficients by

$$c_r = \frac{c_2}{c_1} = \frac{3.5d}{300(1-\varepsilon)\,\mu_0} \tag{3-62}$$

The symbols T_0 and μ_0 designate reference values of temperature and viscosity and the coefficients c_1, c_2 and c_r are thus no longer functions of temperature. The field of velocities in a nonisothermal layer of particles may then be obtained from the solution of the following equation derived from Eq.(3-59)

$$\text{rot}\,\mathbf{G} - \mathbf{G} \times \text{grad}\left[\ln\left(c_1\left(\frac{T}{T_0}\right)^{a+1} + c_2\frac{T}{T_0}G\right)\right] = 0 \tag{3-63}$$

In case of a structurally homogeneous bed, when the inhomogeneity of the velocity field is induced only by its nonisothermality, one can arrange the last equation to the following form

$$\text{rot}\,\mathbf{G} - \left[\frac{(a+1)\left(\frac{T}{T_0}\right)^{a+1} + c_r G}{\left(\frac{T}{T_0}\right)^{a} + c_r G}\right]\mathbf{G} \times \text{grad}\left[\ln\left(\frac{T}{T_0}\right)\right] - \left[\frac{c_r}{\left(\frac{T}{T_0}\right)^{a} + c_r G}\right]\mathbf{G} \times \text{grad}\,(G) = 0 \tag{3-64}$$

Since the coefficient c_r is indirectly proportional to the reference value of the viscosity, the velocity fields resulting from Eq.(3-64) will depend not only on the relative temperature profile but also on the absolute values of temperature. In other words, even in case that the temperature conditions within the layer, expressed in terms of the dimensionless temperature ratio (T/T_0), are the same, the resulting velocity field will be different for different T_0.

If we consider, however, the case of asymptotically fast flows with predominantly inertia dissipation (limiting values of the Reynolds number for this case have been discussed in connection with Eq.(3-33)), we shall obtain the following equation governing the distribution of velocities in a nonisothermal bed passed by an ideal gas

$$\text{rot}\,\mathbf{G} - \mathbf{G} \times \text{grad}\left[\ln\left(\frac{T}{T_0}\right)\right] - \mathbf{G} \times \text{grad}\left[\ln(G)\right] = 0 \tag{3-65}$$

On the contrary, for asymptotically slow flows of an ideal gas through a structurally homogeneous but nonisothermal bed, the following equation holds

$$\text{rot}\,\mathbf{G} - (a + 1)\,\mathbf{G} \times \text{grad}\left[\ln\left(\frac{T}{T_0}\right)\right] = 0 \tag{3-66}$$

For both of these asymptotic cases the governing equations for the velocity field contain only natural logarithms of the dimensionless temperature. The resulting velocity distribution therefore depends on the relative temperature profiles and no longer on absolute values of the temperature.

In order that we may better ascertain the possible extent of thermally induced flow inhomogeneities let us consider the flow through a bed of particles in a cylindrical container while the isotherms have the form of cylindrical surfaces coaxial with the container. We thus deal with the parallel flow under axial symmetry.

Already at this point we have to stress that, in contrast to the case of the parallel flow due to the structural inhomogeneity of the bed, this case of thermally induced inhomogeneity is rather hypothetical.

In case of the parallel flow the only nonzero velocity component of the mass velocity vector is the axial component G_z and the magnitude of the velocity vector, $\mathbf{G}$, thus equals G_z. Under these conditions one can find analytical solution of Eq.(3-64) in the form

$$G_z = -\left(\frac{T}{T_0}\right)^a \frac{1}{2c_r} + \sqrt{\left(\frac{T}{T_0}\right)^{2a} \frac{1}{4c_r^2} + \frac{k_3}{c_r}\left(\frac{T_0}{T}\right)} \tag{3-67}$$

The integration constant k_3 has to be adjusted so as to satisfy the macroscopic balance of mass of the passing gas. Thus

$$G_0 R^2 = \int_0^R 2rG_z \mathrm{d}r \tag{3-68}$$

where G_0 is the mean interstitial mass velocity of gas.

In case of the asymptotically slow flows with predominantly viscous dissipation the velocity profile will be the following

$$G_z = \frac{G_0 R^2}{\displaystyle\int_0^R \frac{2r\mathrm{d}r}{\left(\frac{T}{T_0}\right)^{a+1}}} \cdot \frac{1}{\left(\frac{T}{T_0}\right)^{a+1}} \tag{3-69}$$

while for the fast flows with predominantly inertia dissipation we obtain

$$G_z = \frac{G_0 R^2}{\int_0^R \frac{2r\mathrm{d}r}{\left(\frac{T}{T_0}\right)^{0.5}}} \cdot \frac{1}{\left(\frac{T}{T_0}\right)^{0.5}} \tag{3-70}$$

From here it is thus apparent that for the asymptotically fast flows the deformation of the velocity profile under the parallel flow due to the thermally induced flow inhomogeneities is relatively the same as that due to the structural inhomogeneities of the same extent (compare Eqs.(3-70) and (3-34)). In contrast, the deformation of the velocity profile due to the thermally induced flow inhomogeneities at low velocities is more marked than at the asymptotically fast flow (compare Eqs.(3-69) and (3-70)) and also more marked than the deformation due to the equivalent relative structural inhomogeneity (compare Eqs.(3-69) and (3-36)).

In practical situations, however, the isotherms cannot be parallel to the direction of the flow and one has to solve the complete Eq.(3-64) or (3-66) if the flow approaches these asymptotic states.

If we investigate, for instance, the flow through a cylindrical bed with axial symmetry of the flow the nonzero components of the velocity will be G_r and G_z. Upon introducing again the stream function, this time related to the superficial mass velocity, its definition will be analogous to Eqs.(3-18) and (3-19) with the difference that these expressions relate to corresponding components of the mass and not the volume superficial velocity vector. The equation of continuity under the steady flow, i.e. Eq.(3-52) with the right hand side set equal to zero, is then automatically satisfied and Eq.(3-64) then takes the following form

$$\frac{\partial^2 \psi}{\partial z^2} + \frac{\partial^2 \psi}{\partial r^2} + \frac{\partial \psi}{\partial z}\frac{\partial \ln U}{\partial z} + \frac{\partial \psi}{\partial r}\left(\frac{\partial \ln U}{\partial r} - \frac{1}{r}\right) = 0 \tag{3-71}$$

where U for simplicity designates

$$U = \left(\frac{T}{T_0}\right)^{a+1} + c_{\mathrm{r}}\left(\frac{T}{T_0}\right) G \tag{3-72}$$

The boundary conditions are then formulated on the same tenets mentioned in connection with the flow of an incompressible fluid.

3.4. THE RELATIONSHIP BETWEEN THE INHOMOGENEOUS STRUCTURE OF THE BED AND ENERGY LOSSES OF THE FLOWING FLUID

Solutions of the equations derived for the flow through inhomogeneous layers, obtained mostly by numerical methods, can be found in a number of papers [10,14,16,18,32,33,35,36,38,40,43]. Apart from the possibility of obtaining solutions for the velocity and pressure field in a layer with nonuniform distribution of resistances, a practically important conclusion is that in a number of cases the energy requirements for the flow of the fluid through an inhomogeneous layer, expressed by the pressure differences between the outlet and the inlet cross section, are smaller than those for the flow through a homogeneous layer of equal integral resistance and the same velocity.

This, at first glance rather surprising finding, will be demonstrated in an analytical form for the case of the parallel flow through the layer.

3.4.1. The energy losses in inhomogeneous beds under parallel flow

Parallel flow through the layer occurs, for instance, in a cylindrical bed with axially symmetric inhomogeneity, shown schematically in Fig.3-1, or, generally if the resistance in the bed does not vary axially. The inlet and outlet cross sections are isobaric surfaces and, consequently, the radial velocity components on these surfaces vanish. Zero tangential velocity component is ensured by axial symmetry of the inhomogeneity.

Under the given conditions the velocity vector is identical with its axial component V_z and also the gradient of pressure has a single component in the axial direction. The vectorial differential flow equation (3-1) then degenerates to a single component that can be written in the scalar form as follows

$$\frac{\partial P}{\partial z} = -(f_1 V_z + f_2 V_z^2) \tag{3-73}$$

In case of the asymptotically fast flows with dominating inertia dissipation, one can put $f_1 = 0$ and substitute for V_z from Eq.(3-34) with the following result

$$\frac{\partial P}{\partial z} = -\frac{V_0^2}{(f_2^{-0.5})_0^2} \tag{3-74}$$

The expression in the denominator is the square of the mean value $(f_2^{-0.5})_0^2$, defined in Eq.(3-35). Since under the parallel flow this quantity does not change with axial coordinate, its mean, taken over arbitrary horizontal cross section, is at the same time the mean over the whole volume of the bed.

Let us perform now integration of Eq.(3-74) between the inlet and outlet cross sections over the depth of bed, h, obtaining thus, after some algebraic manipulations, the following dimensionless result

$$\frac{P_{\text{inlet}} - P_{\text{outlet}}}{h\, V_0^2 (f_2)_0} = \frac{(f_2^{-0.5})^{-2}}{(f_2)_0} \tag{3-75}$$

In this expression we have introduced the integral mean of the resistance, (f_2), averaged over the whole volume of the bed as follows

$$(f_2)_0 = \frac{2}{hR^2} \int_0^h \mathrm{d}z \int_0^R f_2\, r\, \mathrm{d}r \tag{3-76}$$

In the case of the parallel flow, of course, the resistance does not vary along the axial coordinate which simplifies the second integration.

By analysis of the results in Eq.(3-75) we come to the conclusion that for a homogeneous layer, in which the resistance is the same in the whole volume, the expression on the left hand side (the dimensionless pressure loss per unit height) equals unity.

On the other hand, from the analysis of properties of the power law functions it follows that the expression on the right hand side of the same equation is less than, or, at most, equal unity. From this we can conclude that the pressure drop across a layer, permitting parallel flow, is lower than the pressure drop across a homogeneous layer of the same integral mean resistance $(f_2)_0$ at the same mean velocity of the flow.

Upon substituting the velocity profile (3-34) valid for the parallel and asymptotically fast flows into the corresponding asymptotic form of Eq.(3-29) for the rate of energy dissipation per unit height of the bed (i.e. setting $f_1 = 0$) we obtain after some arrangement

$$\frac{E}{(f_2)_0\, V_0^3 \pi R} = \frac{(f_2^{-0.5})^{-2}}{(f_2)_0} \tag{3-77}$$

From this expression it follows that the rate of energy dissipation per unit height of the bed under the parallel inhomogeneous flow, in the dimensionless form (3-77) numerically is equal to the dimensionless pressure difference per unit length of the bed, Eq.(3-75), and hence it is lower than the rate of energy dissipation in the homogeneous layer of the same integral mean resistance. Because of the properties of the function f_2, the rate of energy dissipation is the same on any horizontal level.

The rate of energy dissipation in unit volume of the bed, E_v, for the parallel flow and predominantly inertia dissipation, is given as follows

$$E_v = f_2 V_z^3 = \frac{E}{\pi R^2} \tag{3-78}$$

From here we obtain, after substituting the appropriate velocity profile and after some arrangement

$$\frac{E_v}{(f_2)_0 V^3} = f_2^{-0.5} \frac{(f^{-0.5})_0^{-3}}{(f_2)_0} \tag{3-79}$$

Since under the parallel flow f_2 is radially variable, the last equation leads to the conclusion that the rate of energy dissipation related to a unit volume, is also radially variable. In addition, one can state that the distribution of velocities in the layer is such that higher dissipation rates occur in those parts of the column cross section where the local resistance is less than the mean $(f_2)_0$. On the other hand, in parts with higher than the mean resistance the local dissipation rate is lower.

Qualitatively the same conclusions can be arrived at also for the asymptotically slow flow and predominantly viscous dissipation.

Also of practical interest are inhomogeneous structures where the flow is not even approximately parallel.

3.4.2. Change of the flow direction from parallel flow

Equation (3-1) permits one to examine not only the flow through inhomogeneous structures exhibiting different flow resistance in different regions. One can also calculate the velocity and pressure field in a bed where the feed of the fluid does not provide for the parallel and uniform flow although the structure of the bed is uniform.

An example of this type of the flow was investigated by Radestock and Jeschar [73,75]. These authors considered a cylindrical reactor fed by gas through a slot along the circumference of the reactor near its bottom. Although the bed is uniform the gas has gradually to change its direction from the horizontal one directed toward the center of the reactor to the vertical parallel uniform flow in the direction of the reactor axis.

A sketch of the streamlines and isobars is shown in Fig.3-4. Since the height at which the gas flows essentially uniformly is not known in advance, several test calculations have to be run in order to find with sufficient accuracy at which distance one can apply the boundary conditions stipulating constant pressure over the whole cross section.

The calculations furnished the following general conclusions. The flow is essentially uniform and parallel at a height about one reactor radius above the gas feed. The results, of course, differ somewhat depending on the mean velocity of the flow.

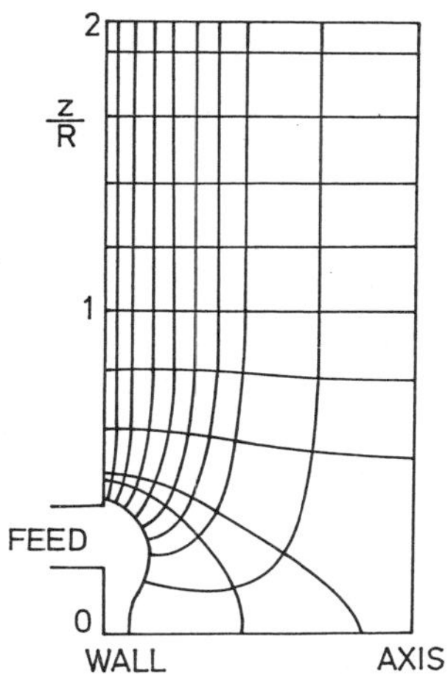

Fig.3-4
Streamline and isobar pattern in a cylindrical bed fed sideways.

For low gas mean velocity the streamlines display greater radii of curvature suggesting a somewhat slower transition to parallel flow. The height of the reactor necessary to reach the same degree of flow uniformity is therefore slightly higher. This case of slow flow through a uniform bed actually corresponds to the "potential flow" and the equipotential lines are the isobars.

At higher mean gas velocities a greater fraction of the overall pressure drop is concentrated near the gas inlet and the region near the axis while the region near the reactor bottom in which low local velocities prevail grows smaller.

In blast furnace operations one also encounters this type of flow and since the gas velocities here are fairly large the compressibility of the flowing gas may be an important factor in contrast to the case of the conventional reactor. With increasing compressibility of the flowing gas the pressure gradients near the inlet will decrease accordingly. The effect of increasing gas velocity on the streamline and the isobar pattern is thus somewhat offset by increasing compressibility as these two effects work against each other.

A similar situation has been analyzed both experimentally and theoretically by Fukutake and Okabe [80]. Their experiments however were carried out on a "two-dimensional" rectangular bed fed horizontally from the side near the bottom. Experiments with spherical particles of different size showed a very good agreement of the computed pressure field with the pressure measurements along the height of the bed.

The flow conditions near the feed can, of course, also be affected by the structure of the bed near the feed. If there is a core of increased resistance on the bottom around the axis of a cylindrical bed fed from the side near the bottom the streamlines tend to bypass the high resistance core in such a way that part of the gas flow turns abruptly upwards before hitting the high resistance core, while another substantial part passes almost horizontally, taking the shortest path across the high resistance core and turning upwards again at the far side of the reactor in the low resistance region. This situation has been investigated by Szekely and Poveromo [95].

As found by Jeschar et al [150] radial supply of gas into a bed in which there is already a uniform flow of gas from below causes the gas streaming from below to be pushed away from the wall toward the center. The overall direction of the flow turns gradually to purely vertical again while the required height to reach parallel flow is approximately equal to the bed radius. Cross sectional areas of the bed available for the gas coming from below and that fed from the side are proportional to the flow rates of the two respective streams.

In addition, the gas fed from the side flows exclusively in the peripheral region while the gas coming from below exclusively in the central region. The mixing of the two streams is due only to "diffusion".

A similar analysis was made by the same authors [150] of the case in which there are two side streams into a bed passed by gas. Some of the gas is being withdrawn from the bed via the lower side stream while some of the gas is being fed into the bed through the upper side stream. It was found that when the flow rate of gas withdrawal via the lower side stream was equal to or higher than the flow rate of gas coming from the bottom, no bottom gas penetrated into upper part of the bed to mix with the gas fed via the upper side stream. Also, no gas fed via the upper side stream returned into the bottom part to be withdrawn via the lower side stream if the volume flow rate in the bed above the upper inlet was higher or equal the flow rate of gas fed via the upper side stream.

Thus no mixing of gas coming from below with the gas fed by the upper side stream takes place if the volume flow rate of gas coming from below equals the flow rate of gas withdrawal by the lower side stream and if the volume flow rate of gas fed by the upper side stream equals the flow rate in the bed above the upper side stream.

The zone around gas withdrawal in the lower part of the bed and the zone around gas feed in the upper part of the bed are separated by a region of relatively weak flow if the distance of the feed and withdrawal side stream is sufficiently large. In this case the conditions for "diffusional" mixing are poor. This picture, of course, changes as the two side streams come closer together. The region between the feed and the withdrawal level is characterized by high local velocities of opposite direction in the upper and the lower part. This naturally gives sample opportunities for intensive "diffusional" mixing of gas between the two regions.

3.4.3. The flow in constrictions

Nonuniformities of the flow appear not only as a consequence of the spatially variable resistance but also due to the obstacles in the flow path. A scheme of a certain type of the obstacle representing actually a constricition of the cross section is shown in Fig.3-5. This type of the flow obstacle has been investigated by Berninger and Vortmeyer [127,128] both theoretically and experimentally.

In practice the obstacle may be a cooling fin, a thermocouple well or a clogged part of the catalyst bed due to the deposits of the fines, coking, polymerization or clustering of particles.

The scheme in Fig.3-5 shows a cylindrical bed whose depth is considered sufficently large so as to have a uniform flow in the inlet and outlet surfaces. Equally one can consider an infinite bed and place the condition of the uniform flow at a finite distance satisfying this requirement. At the level $z = 0$ the cross section of the bed is constricted from the radius R of the bed to a radius R_1 .

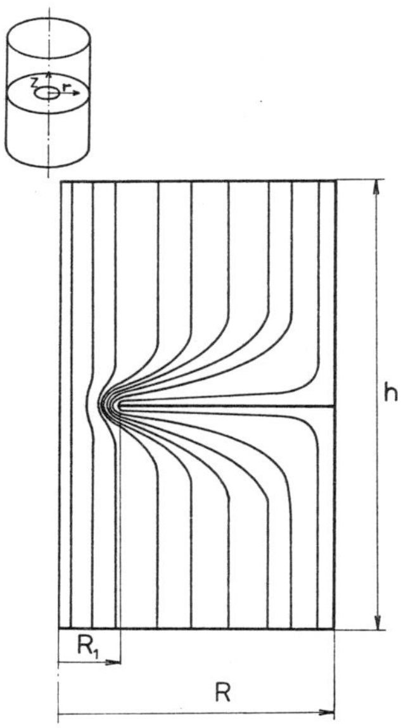

Fig.3-5
Streamline pattern in a cylindrical bed with constriction.

Bernier and Vortmeyer [127] solved this problem using Eq. (3-17) in the two-dimensional form. The conclusions arrived at from a number of case studies were summarized as follows:

Equalization of the velocity profile takes place within a short flow path. No vortices appear in the vicinity of the constriction. The computed results are symmetric with respect to the plane passing through the constricting annulus. With increasing velocity the computed results do not change significantly but the equalization of the velocity takes place slightly faster. At equal Reynolds numbers based on particle size the flow pattern is also similar and the particle size does not exert any appreciable effect.

The streamlines shown in Fig.3-5 are drawn only approximately on the basis of the velocity field published by Berninger and Vortmeyer [127]. These authors provided also experimental confirmation of the computed velocity field by photographing the spread of a dye in the bed at various time instants. The agreement was remarkably good considering that the calculations did not take into account the effect of the presence of the constricting annulus on the void fraction in its neighbourhood.

Analysis of the flow problems of this type, using Eq. (3-11), i.e. the equation with eliminated pressure, yields the following conclusions: In the case of an incompressible fluid the problem is symmetric with respect to the plane passing through the constricting annulus. It suffices then to solve the streamline field either above or below the constriction. The boundary condition directly within the constriction is easily formulated on the grounds that the velocity vector is perpendicular to the plane of constriction and the radial velocity component here therefore vanishes.

Further we can see that for low mean flow velocity, when the first of the Ergun terms, Eq.(3-2), becomes dominant, corresponding equations for the velocity (3-11) and pressure (3-17) may be modified to the Laplace equations for the stream function and pressure. Their solution in terms of the dimensionless velocity and pressure, using the mean velocity, mean resistance of the bed (in this case equal to the uniform local resistance of the bed) and the bed radius as scaling factors, is then independent (for the geometry given by the simplexes h/R, R_1/R) of the velocity and resistance. Pressure drops for a different resistance of the bed, due to e.g. different particle size, are then easily obtained by multiplying the dimensionless pressure drop by the corresponding resistance.Also for the other extreme, when the quadratic Ergun coefficient, Eq.(3-3), dominates, one can make the same conclusions regarding the dependence of the results on the mean velocity and the mean resistance of the bed $(f_2)_0$ (again equal to the uniform resistance of the bed). In contrast, however, the underlying asymptotic equations are no longer the linear Laplace equations and the dimensional pressure drop across the constriction is now proportional to the square of the mean velocity V_0 .

In the transition region, approximately between the Reynolds number 1.5 and 150, where both terms of the Ergun equation are significant, the solution of the velocity as well as pressure equations, even in terms of dimensionless variables, depend on the mean velocity and the solution has to be obtained individually for each value of interest.

Naturally, the results of the calculation depend also on the geometry of the constriction, namely the simplex R_1/R. This effect can be characterized qualitatively using the results of Berninger and Vortmeyer [127] and Fig.3-5, as follows. The region where the velocity vector, or the streamlines significantly deviate from the largely vertical direction forms in the r-z plane of coordinates an isosceles triangle with its base on the wall with the constriction being its height. Thus, of course, for large values of R_1/R, leaving less free cross section for the flow, the height of the bed affected by nonparallel flow will be greater. As a rule of thumb one could say that doubled constriction ratio R_1/R will cause doubling of the height of the bed affected by the nonuniform flow.

3.4.4. The energy losses and flow maldistribution in beds with axial symmetry of structural inhomogeneity

Let us consider again a cylindrical bed as in Fig.3-1 divided by a hypothetical cylindrical surface of radius $0.7R$. The height of the bed is taken to be $2R$. This situation is sketched once more in the left hand upper corner of Fig.3-6 and may represent part of the shaft of a blast furnace [18,34]. The left hand vertical edges of the sketches in Fig.3-6 represent always the axis of the shaft, while the right hand edge the wall.

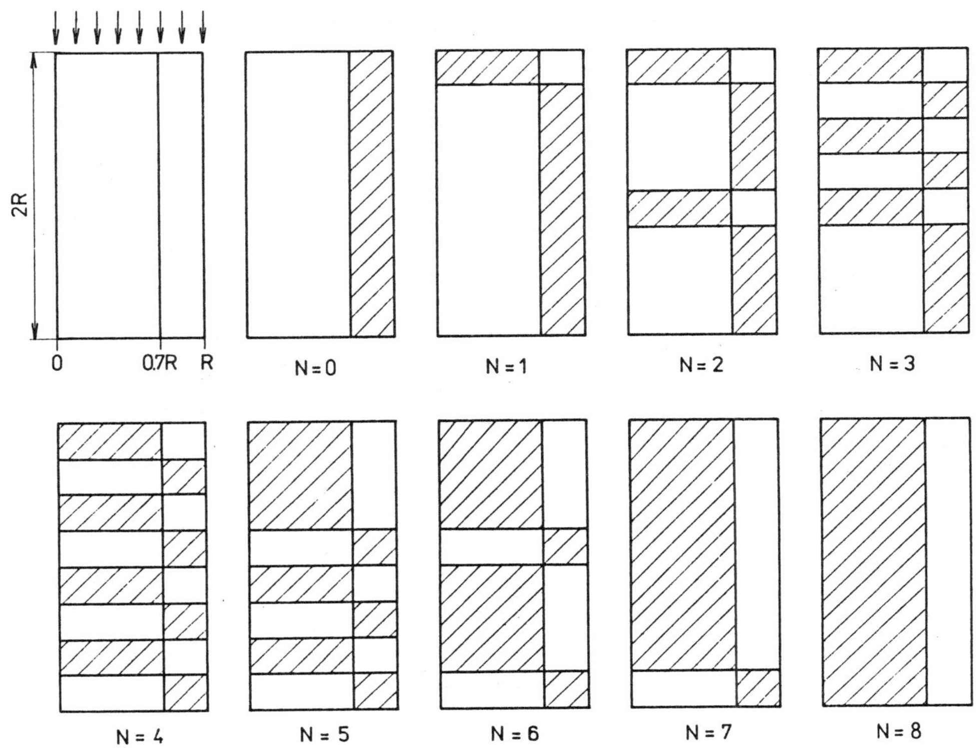

Fig.3-6
Sketch of the bed divided by a hypothetical coaxial cylindrical surface and seven horizontal hypothetical plates into sixteen compartments. Shaded and empty areas distinguish compartments occupied by different particles.

Further we shall assume that the bed is divided by hypothetical horizontal planes into 8 sections of equal height $R/4$. The bed thus consists of 16 hypothetical compartments charged by two kinds of particles differing in size given by diameters d_c and d_a (say coke and agglomerate in case of a blast furnace).

First we shall assume that agglomerate (d_a) occupies the cylindrical space around the axis while coke (d_c) occupies the annular space between the hypothetical cylindrical surface and the wall. This situation is designated in Fig.3-6 as configuration "0". In this and all following configurations the volumes occupied by agglomerate/coke are shown by empty/shaded areas.

The following configuration "1" is obtained from the previous one by switching the coke and agglomerate in the top most hypothetical compartments. To get configuration "2" we switch, in addition, coke and agglomerate in the compartments on the fifth level counted from the top. The switching procedure is continued through configurations "3" to "8" until an entirely reversed configuration, compared to the configuration "0", is reached. Namely, agglomerate in configuration "8" occupies the annular space adjacent to the wall of the shaft. Of particular interest is the configuration "4", characterized by regularly alternating structure on individual horizontal levels.

Since the particles of coke, d_c, are larger, the shaded areas in sketches in Fig.3-6 represent regions of smaller resistance to the flow. As the volumes of the cylindrical and the annular part of the shaft differ only little the integral mean resistance of individual configurations varies only little.

In practice, of course, the height of the shaft is much higher than $2R$ (as adopted here), usually 4-5 times the shaft diameter, and this also plays an important role.

The flow equations were solved [18,34] for an incompressible gas and asymptotically fast flows satisfying condition (3-20). This was done not with the aim to simplify the underlying problem, but rather for the sake of generalization of the results obtained. Under these conditions the results become independent of absolute pressure, which appears only as the "background" pressure on the

Fig.3-7
Streamline pattern for the flow through a bed shown in Fig. 3-6 as N=4 for d_c/d_a = 50/5.3 [mm/mm]. The left hand side depicts the axis (identical with streamline 0); the right hand side depicts wall (identical with streamline 0.5). The curves from right to left are streamlines 0.4; 0.3; 0.2; 0.1; 0.08; 0.06; 0.04; 0.02; 0.008; 0.006; 0.004; 0.002.

isobar pattern, and the velocity field in the dimensionless form is independent of the mean velocity of the flow, V_o. The corresponding equation in terms of the stream function is thus the one written in Eq.3-21.

The pattern of streamlines obtained for the structure $N = 4$ is shown in Fig.3-7 and $d_c/d_a = 50/5.3$ and it may be shown that the result does not depend on the absolute values of d_c and d_a but their ratio only. The streamlines exhibit a typical "wavy" shape as the gas tends to pass through regions of lower resistance.

The streamline pattern for other configurations, N, will be characterized subsequently in an integral form of a criterion characterizing the nonuniformity of the flow.

It is noted that for the configurations $N = 0$ and 8 the corresponding flow is parallel and the flow distribution may be therefore obtained analytically from Eq.(3-34). This formula was used to check the results obtained numerically.

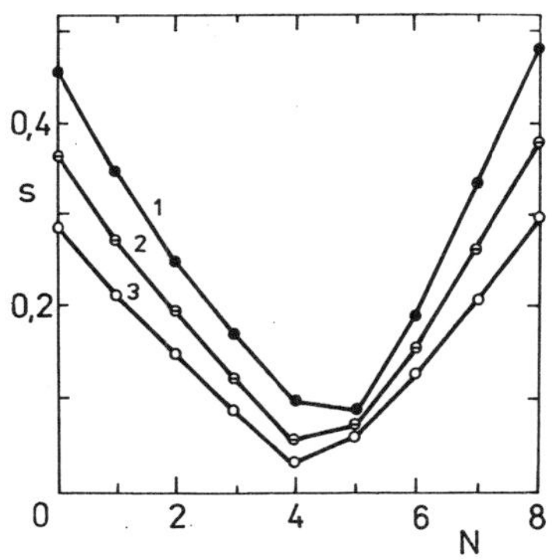

Fig.3-8
Criterion of flow nonuniformity for structures N shown in Fig.3-6 and various particle diameters. Curve 1 — $d_c/d_a = 50/5.3$; curve 2 — $d_c/d_a = 40/7$; curve 3 — $d_c/d_a = 35/9$ [mm/mm].

Fig.3-8 plots the nonuniformity of the flowcriterion, defined as

$$s = \frac{1}{V_0}\left\{\frac{1}{R}\int_0^R \left[V_{z0} - \frac{1}{R}\int_0^R V_{z0}\,\mathrm{d}r\right]^2 \mathrm{d}r\right\}^{0.5} \tag{3-80}$$

where V_{z0} designates the axial velocity component averaged for the given radial coordinate along the height of the shaft:

$$V_{z0} = \frac{1}{h}\int_0^h V_z \mathrm{d}z \tag{3-81}$$

The value of s is zero for perfectly uniform flow and increases with increasing flow maldistribution. Values of s are plotted in Fig.3-8 against the number of the structure, N. The results are shown for three ratios of coke and particle diameters d_c/d_a .

From the last figure it is apparent that the configuration $N = 4$ with regularly alternating layers exhibits the best flow uniformity expressed by the criterion s. Flow maldistribution is expectedly the most severe for the two configurations providing for the parallel flow ($N = 0$ and 8) and the uniformity of the flow improves significantly as one proceeds from both ends toward the regularly alternating configuration $N = 4$.

Due to the character of cylindrical coordinates and slightly varying mean integral resistance of the bed the curves in Fig.3-8 are not symmetrical with respect to the configuration $N = 4$. As could be expected the maldistribution of the resulting flow decreases for all configurations with decreasing disparity of the coke and agglomerate particles, i.e. with the d_c/d_a ratio.

The deviations from the flow uniformity are always of such a nature that the axial velocity component averaged along the height of the bed, V_{z0}, for various radii is higher on that side of the surface separating the central and peripheral compartments, where there is greater fraction of the larger (coke) particles. To be more definite, greater mean axial velocities exist for radii greater than 0.7 R for configurations $N = 0$ to 3 and vice versa.

The calculations [34] yielded also the pressure distributions for all configurations and were used to construct the isobar pattern. The equation employed for the calculation was Eq.(3-16) written in cylindrical coordinates with axial symmetry and the simplification given by Eq.(3-20). The boundary conditions stipulated that the inlet and outlet are isobaric surfaces. We note that once the velocity distribution is known the whole right hand side of Eq.(3-16) is a known function of spatial coordinates and the calculation poses no numerical difficulties.

In contrast to Eq.(3-11), the resistance coefficients f_1, f_2 do not appear in the equation for pressure distribution (3-16) as logarithms. Consequently, the dimensional pressure patterns depends not only on the d_c/d_a ratio but on the absolute values of d_c and d_a as well. The effect of absolute values of d_c and d_a, however, can be assessed relatively easily because the obtained results for various d_c, d_a but the same d_c/d_a ratio differ only by a constant coefficient which may be incorporated into the dimensionless definition of the mean resistance of the bed.

On the other hand, the effect of d_c/d_a ratio, keeping the same integral mean resistance of the bed, can be assessed from the results obtained in terms of the dimensionless pressure drop.

The constructed isobar pattern for the configuration $N = 4$ and $d_c/d_a = 50/5.3$ is depicted in Fig.3-9 (d_c and d_a in milimeters). These isobars thus correspond to the flow pattern shown in Fig.3-7. As before, the left hand edge of the figure is the axis of the shaft while the right hand edge represents the wall. The values assigned to the isobars are dimensionless pressures $P/(h(f_2)_0\ V_0^2)$. The pressure at the outlet surface (bottom of the figure) was assigned arbitrarily a zero value (zero background pressure), while the value for the inlet surface (top of the figure) indicates the dimensionless pressure drop per unit height of the bed.

The obtained isobars tend to crowd into regions occupied by agglomerate. Higher density of the isobars here suggests higher local pressure gradients and hence higher local pressure losses.

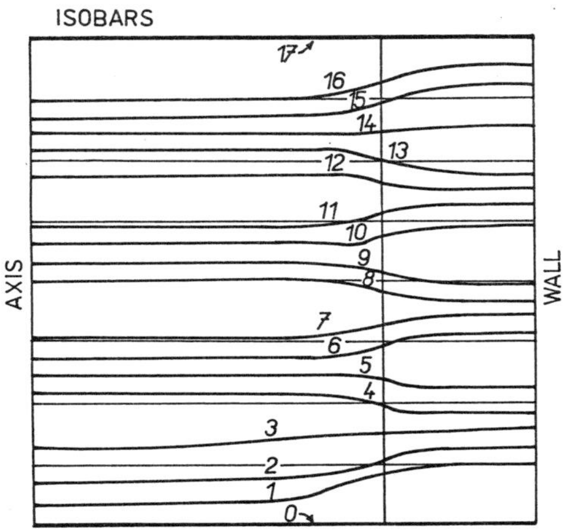

Fig.3-9
Isobar pattern for the structure shown in Fig.3-6 as N=4 and d_c/d_a = 50/5.3 [mm/mm]. The curves 1 - 17 represent isobars 0.05; 0.1; 0.15; 0.20; 0.25; 0.30; 0.35; 0.40; 0.45; 0.50; 0.55; 0.60; 0.65; 0.70; 0.75; 0.80; 0.8485 of the dimensionless pressure ΔP^* .

Another fact clearly manifested by Fig.3-9 is that in inhomogeneous structures the isobars are not horizontal planes. Pressure measurements realized through ports in the wall (right vertical edge of the figure) thus reflect pressure conditions corresponding to the flow distribution as affected by the structure of the bed in the vicinity of the wall. These conditions need not be typical, as in the examined case, for the bulk of the bed.

Fig.3-10 is a plot of the dimensionless pressure drop per unit height of the bed defined as

$$\Delta P^* = \frac{\Delta P}{h V_0^2 \, (f_2)_0} \tag{3-82}$$

versus the configuration number N. From the definition of the dimensionless pressure drop it follows that the numerical value of the dimensionless pressure drop per unit height of the bed of the same mean integral resistance $(f_2)_0$ and the same velocity V_0 such that it satisfies the conditions of asymptotically fast flow, Eq.(3-20), equals unity. From the figure it is apparent that all investigated configurations exhibited for all investigated particle sizes pressure losses lower than unity. The pressure losses at the nonuniform flow were thus lower than those of the uniform flow of equal mean resistance $(f_2)_0$ and at equal mean velocity V_0 .

From Fig.3-10 it further follows that the pressure losses in the dimensionless form decrease with increasing disparity of the coke and agglomerate particles, i.e. with increasing d_c/d_a ratio. It may be noticed however that the differences between various particle disparities but the same configuration number N diminish with improving homogeneity of the flow. For the regularly alternating configuration $N = 4$ the results for different d_c/d_a ratios are mutually the nearest.

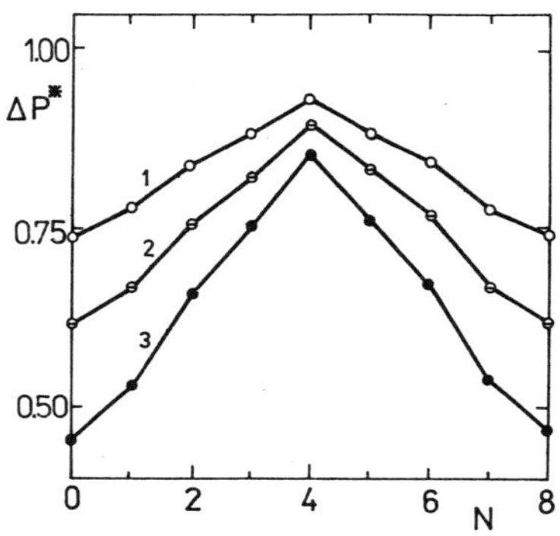

Fig.3-10
Dimensionless pressure drop for structures N shown in Fig.3-6 and various particle diameters. Curve 1 — $d_c/d_a = 35/9$; curve 2 — $d_c/d_a = 40/7$; curve 3 — $d_c/d_a = 50/5.3$ [mm/mm].

It is stressed though that the numerical results plotted in Fig.3-10 are valid for the given absolute values of particle diameters d_c, d_a and not generally for the given d_c/d_a ratio. This can be proved as follows: Let us designate the volume fraction of the bed occupied by coke particles by X. The integral mean resistance is then given by

$$(f_2)_0 = k\frac{1}{d_c}\left(X + (1 - X)\frac{d_c}{d_a}\right) = k\frac{1}{d_0} \tag{3-83}$$

where d_0 designates the mean particle diameter defined as

$$\frac{1}{d_0} = \frac{X}{d_c} + \frac{1 - X}{d_a} \tag{3-84}$$

and the meaning of the constant k can be found easily from comparison with Eq.(3-3). In our investigated case X is nearly independent of the configuration number N and equals approximately 0.5. Local dimensionless resistances

$f_2/(f_2)_0$, where f_2 is a function of r,z, thus assume the following values: in the region occupied by coke:

$$\frac{1}{X + (1 - X)\dfrac{d_c}{d_a}}$$

and in the region occupied by agglomerate:

$$\frac{\dfrac{d_c}{d_a}}{X + (1 - X)\dfrac{d_c}{d_a}}$$

The field of the dimensionless local resistances and consequently also the dimensionless pressure drop are thus functions of the d_c/d_a ratio and of the fraction X but not of the absolute values of particle diameters.

The picture of the relationship between the pressure losses, the configuration of the bed and the size of particles forming the bed will be different if we examine the dimensional pressure losses. Dimensional pressure losses per unit height of the bed are given by

$$\frac{\Delta P}{h} = \Delta P^* \, (f_2)_0 \, V_0^2 = \Delta P^* \, \frac{k}{d_0} \, V_0^2 \tag{3-85}$$

Fig.3-11 plots the quantity $\Delta P^* \, (f_2)_0/k$ proportional for given V_0 according to the last equation to the dimensional pressure drop per unit height of the bed, i.e. $\Delta P/(hV_0^2 k)$. For the majority of structural configurations of the inhomogeneity, N, the pressure drop increases with the d_c/d_a ratio, i.e. with increasing particle disparity.

For parallel flows through configurations $N = 0$ and 8, on the other hand, the dimensional pressure losses are minimal for the largest d_c/d_a ratio, while for the configurations $N = 1$ and 7 the dimensional pressure losses depend little on the disparity of particle diameters.

Since for all configurations the mean resistance of the bed is nearly the same (only small variation of X from 0.51 to 0.49 for configurations 0 to 8) it can be concluded that the configuration (structure) of the bed is the major factor in increasing the dimensional pressure losses for higher d_c/d_a ratios.

Structures inducing higher flow uniformity force the gas to flow through regions of higher resistance with necessarily higher energy losses. Also, the same structures tend to create more complicated streamline patterns with higher degree of the "cross" flow, longer path and corresponding excess pressure losses.

Numerically, the results plotted in Fig.3-11 are valid only for the given d_c/d_a ratio and the mean particle size d_0. Results for the same disparity ratio but different mean particle size can be obtained easily by scaling the results by the appropriate mean particle diameter as indicated by Eq.(3-83).

The inverse problem of finding the effect of different d_c/d_a ratio while keeping the mean resistance of the bed (d_0) constant is answered through the solution in terms of the dimensionless pressure drop.

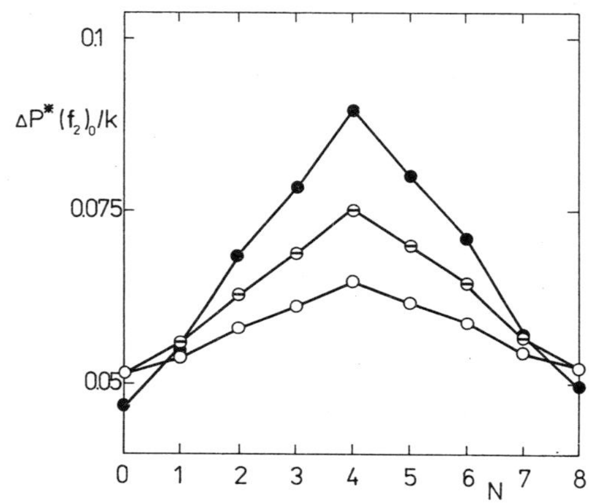

Fig.3-11
Quantity proportional to dimensional pressure drop for structures N shown in Fig.3-6 and three particle diameter ratios. Empty circles - $d_c/d_a = 35/9$; halved circles - $d_c/d_a = 40/7$; full circles - $d_c/d_a = 50/5.3$ [mm/mm].

In general it can be said that in configurations permitting higher extent of the cross flow the reduction of the pressure losses through greater disparity of particle diameters grows weaker. It is stressed, however, that numerical results of the underlying calculations depend also on the simplex h/R, taken here to be equal 2. The effect of this parameter is quite important for quantitative conclusions and it will be examined separately for the configuration $N = 4$.

The possibility of reducing the energy requirements for the flow has always to be considered jointly with the associated flow maldistribution. Both factors, as was demonstrated in Figs.3-8 and 3-11, work against each other and mutual counterproductivity of both effects is the stronger the greater the d_c/d_a ratio.

Charging of the blast furnace often takes place with the aid of a bell closing the furnace and diverting at the same time the falling particles toward the wall. The "interfaces" between the principal components of the charge, coke and agglomerate, are then essentially conical surfaces slanting toward the center of the furnace, see e.g. Jeschar et al [151]. The inclination of the interfaces varies according to the properties of the particles, angle of repose and the sequence in which they are charged. As the charge descends in the course of the process down the shaft the inclination of these interfaces also varies somewhat.

The structure given rise to by this charging practice [39,40] is shown schematically in Fig.3-12. This figure shows three structures designated as A,B,C. In all three cases the sketch shows the part of the shaft filled by two consecutive chargings and it is assumed that the structure as sketched persists throughout the height of the furnace. The shaded and empty areas distinguish regions occupied by the particles charged first and second, i.e. the coke and agglomerate. Common to all three structures is that the volumes occupied by particles of diameter d_1 and d_2 are in all cases the same and mutually equal.

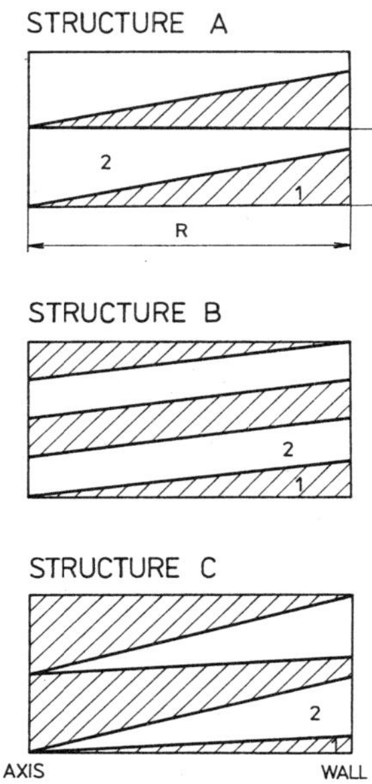

Fig.3-12
Three structures simulating a blast furnace charge. Shaded and empty areas distinguish regions occupied by different particles.

The depth of the layer of particles charged first shown by the shaded area, is equal near the wall to 3/4 of the depth of the layer of a single charge, h. Since this depth at the axis of the bed is equal zero it is seen that the structure A places particles charged first predominantly to the wall region of the furnace.

In the case of the structure B the depth of the layer of particles charged first near the wall and on the axis is the same and equal to 1/2 of the depth of the layer of a single charge.

Structure C places the major portion of the particles charged first into the axial region of the furnace. The depth of the layer of particles charged first near the wall is equal to 1/4 of the depth of the layer of a single charge.

Flow of gas through the structures just described has been investigated [39,40] by solving the corresponding equation of motion in cylindrical coordinates with axial symmetry, Eq.(3-21). Since we have considered that these structures repeat periodically along the vertical coordinates, the top and bottom surface of the two layers of a single charge, shown in Fig.3-12, are not generally isobaric surfaces.

The boundary condition utilized the periodicity of the solution by substituting the results, found at the horizontal level in the middle of the structure, to the top and bottom horizontal surfaces. The obtained results of the numerical solution depend naturally on the size of the two kinds of particles forming the structures A,B and C, on the sequence of charging (which particles are charged first) and on the ratio of the depth of the layer of a single charge to the radius of the furnace R.

The obtained numerical results of the solution in terms of various characteristics are summarized in Table 3-1. In general one can, however, say that for the investigated values of the dimensionless depth of the layer of a single charge, which are generally small compared to the diameter of the shaft, the flow of gas is unable to respond sufficiently rapidly to the vertically variable resistance to the flow and the streamlines therefore are nearly vertical straight lines. To be more precise, the streamlines exhibit a slightly wavy shape. The flow is thus nearly parallel but not uniform as will be apparent from the following. Corresponding maps of the radial velocity component exhibit maximum absolute values along the "interfacial surfaces" separating regions occupied by particles of different size. Isobars indicate nonuniform distribution of the forces exerted by the flowing gas and, as already mentioned, the top and bottom surfaces of the structures A, B, C are not isobaric surfaces. The pressure drop per single layer of a charge thus varies along the radius of the shaft.

Table 3-1 summarizes results of 12 runs for the three structures (see the second column) and for two values of the dimensionless height of the layer of a single charge (the third column) and particle sizes shown in the fourth column (d_1 in the numerator being charged first).

The fifth column of this table shows the deviation, s, Eq. (3- 80), of the velocity V_{z0}, Eq.(3-81), from its mean, the dimensionless mean being shown in the sixth column. From these data, characterizing the uniformity of the flow it is seen that runs No. 2,5,8 and 11 all exhibit small flow nonuniformity.

This is so because all these runs pertain to the structure B where the radial nonuniformity of the resistance to the flow is small. This is again manifest through the results in the seventh column of this table showing the ratio of the mean axial velocity component near the wall and on the axis. In all cases this characteristic for structure B is close to unity.

Structures A and C, depending on the sequence of charging, bring about either strongly peripheral or axial flow. For instance, for run No.1, i.e. the structure A with the smaller particles being charged first, the flow in the peripheral region of the shaft amounts in terms of the mean axial velocity (the seventh column) to only slightly more than 50% of the flow in the axial region. For run No.4, i.e. the same structure but the reverse sequence of charging, when larger

Table 3-1

Run No	Structure	$\frac{h}{R}$	d_1/d_2 [mm/mm]	Standard deviation of V_{z0} from the mean: s	mean V_z/V_0
1	2	3	4	5	6
1	A	0.36	7/40	0.233 8	1.131 09
2	B	0.36	7/40	0.001 7	1.000 62
3	C	0.36	7/40	0.138 6	0.920 79
4	A	0.36	40/7	0.132 9	0.924 21
5	B	0.36	40/7	0.002 1	1.000 68
6	C	0.36	40/7	0.243 3	1.136 08
7	A	0.72	7/40	0.222 1	1.126 14
8	B	0.72	7/40	0.008 0	1.004 02
9	C	0.72	7/40	0.152 7	0.912 51
10	A	0.72	40/7	0.139 9	0.919 90
11	B	0.72	40/7	0.008 5	1.004 04
12	C	0.72	40/7	0.245 1	1.139 06

Table 3-1 – continuation

$\frac{(V_{z0})(r = R)}{(V_{z0})(r = 0)}$	ΔP^*	$\frac{\Delta P^*(r = R/2)}{\Delta P^*_{mean}}$	$\frac{\Delta P^*(r = R)}{\Delta P^*_{mean}}$	$\frac{\Delta P^*(r = 0)}{\Delta P^*_{mean}}$	$\frac{((f_2)^{0.5})_0^{-2}}{(f_2)_0}$
7	8	9	10	11	12
0.502 3	0.909 7	1.003 4	1.011 2	0.967 5	0.850 6
0.997 0	0.972 3	1.000 3	1.001 5	0.996 0	0.854 9
1.649 1	0.915 9	1.003 4	1.011 6	0.996 7	0.854 0
1.598 7	0.938 9	1.000 9	1.003 7	0.990 0	0.849 0
0.992 4	0.977 5	1.000 1	1.001 3	0.997 0	0.855 4
0.485 7	0.903 5	0.997 5	0.992 0	1.024 0	0.856 0
0.515 7	0.880 8	1.004 8	1.015 9	0.954 2	0.850 8
0.973 7	0.932 9	1.000 7	1.003 1	0.991 2	0.854 9
1.762 4	0.870 6	1.005 7	1.019 2	0.944 7	0.854 0
1.646 5	0.915 5	1.001 2	1.004 2	0.988 3	0.849 0
0.972 1	0.939 6	1.000 5	1.002 1	0.994 0	0.855 5
0.486 0	0.891 8	0.995 2	0.984 7	1.046 7	0.856 0

particles are predominantly in the peripheral region, the mean axial velocity exceeds that in the axial region by nearly 60%.

The depth of the layer of a single charge appears to have only a minor effect on the computed results of flow nonuniformity as may be observed from comparison of runs 1-6 with runs 7-12. Expectedly, higher relative depth of a single charge causes stronger flow nonuniformity but the differences are not large in the region of values investigated.

The eighth column in Table 3-1 presents dimensionless pressure drops per unit height in the form when the pressure drop of a homogeneous layer of equal integral mean resistance $(f_2)_0$ is equal to unity. From the data in this column it is seen that corresponding pressure drops are in all cases lower while the energy savings for the structure B are generally smaller than for the structures A and C. This is associated with the lower nonuniformity of the flow induced by the structure B.

The pressure drop data may be compared with the quantity computed in the twelfth column which expresses the pressure drop if the flow on each horizontal level had been immediately redistributed in accord with the local distribution of the resistances. From this column it is seen that these hypothetical minimal pressure drops differ little for individual runs. On the other hand the computed reductions of the pressure drops for the computed patterns of the flow account for a sizeable portion of energy savings at least for the structures A and C (compare the eighth and the twelfth column).

Columns 9-11 of the table characterize the nonuniformity of the pressure field. These data represent dimensionless pressure drops per unit height in various radial positions, all scaled by the mean pressure drop. On these values it is interesting to note that pressure drops for $r = R/2$ (where it is technically not feasible to measure it) correspond best to the mean pressure drop of the bed. Pressure drops in the wall region, where they are actually measured, differ in case of structures A and C by about 1%. In the axis these differences amount to about 3%.

One can conclude that in the problem examined the structure itself and the sequence of charging (which component comes first) have a significant effort on the uniformity of the flow. The differences in the wall and peripheral region may amount to up to 100%. This gives rise to considerable possibilities of affecting the uniformity of the flow by modifications of the charging practice.

The differences in the uniformity of the pressure field and hence the local pressure drop are considerably less conspicious. Nevertheless, pressure measurements through ports in the wall do not give precise information about the pressure field prevailing within the bed especially if the flow nonuniformity is high.

3.4.5. Energy losses and flow maldistribution in beds with angular distribution of structural inhomogeneity

In the paragraph dealing with parallel flow we have already investigated the bed where the resistance varied with the angular coordinate, but not with the radial and axial coordinate. It is interesting to investigate this case after removing the simplification that the resistance does not vary with the axial coordinate.

Answers to such questions are important not only theoretically but also practically. This is so because modern blast furnaces utilize the so-called bell-less devices for charging the furnace. These devices are essentially computer controlled troughs permitting deposition of a given component (coke, ore) into specific locations. This further enhances the potential of utilizing structural measures aimed at improvement of the regime of the technological process.

Let us consider again [36] a bed filled alternately with two kinds of particles of different diameter d_c, d_a. First we charge particles of diameter d_c and the depth of the first deposited layer, h_{c1}, will vary with the angular coordinate as follows

$$h_{c1} = a + b \cos \theta \tag{3-86}$$

The next layer charged into the bed would be composed of particles of diameter d_a and its depth, h_{a2}, will be the following function of the angular coordinate

$$h_{a2} = a + b \cos (\theta - \gamma) \tag{3-87}$$

This means that the second layer is angularly shifted by an angle γ with respect to the previous layer. The third layer is composed again of the particles of diameter d_c but this layer is not angularly shifted with respect to the second layer.

$$h_{c3} = a + b \cos (\theta - \gamma) \tag{3-88}$$

This procedure is repeated so that each odd layer consists of d_c sized particles, while each even layer consists of d_a sized particles. In addition, every even layer is shifted angularly by the angle γ.

Simple geometrical considerations show that the mean depth of a single charge is "a" and upon taking $\gamma = 3\pi/4$ the surface of the sixteenth layer will again be a plane. The total depth of the bed is $h = 16a$, while each of the two components occupies one half of the volume of the bed. In addition the resistance of the bed, averaged along the height, is independent of the coordinate θ.

The calculations for this type of structural inhomogeneity were carried out [36] for

$$a = 0.25R$$

$$b = 0.04R$$

It can be shown [35] that for this type of structural inhomogeneity, when the resistance varies with the coordiante z and θ only, the nonvanishing velocity components will be V_z and V_θ. Their spatial distribution can be obtained from the solution of Eq.(3-14a), while both components must satisfy the continuity equation.

We note that Eq.(3-14a) still contains the coordinate of radius although there is no radial flow. It, however, appears only as a parameter of the solution.

Formally it could be removed by a suitable transformation of the coordinate of angle, replacing θ e.g. by $(r\theta)$. Nevertheless, the radius would still enter the solution implicitly through the radial dependence of the upper limit of integration for the new coordinate. Alternatively, one can use dimensionless coordinates and the radius, determining the cylindrical surface for which the solution is obtained, enters in the form of a simplex (R/r).

Equation (3-14a) was solved [36] by introducing first the stream function in the usual dimensionless form as

$$V_\theta/V_0 = -2\pi r \frac{\partial \psi}{\partial z} \tag{3-89}$$

$$V_z/V_0 = 2\pi \frac{\partial \psi}{\partial \theta} \tag{3-90}$$

and upon considering that the inlet and the outlet surface of the bed are isobaric surfaces. The derivative of the stream function with respect to the axial coordinate is thus zero and for continuity reasons we can choose for the inlet surface that

$$\psi = 0 \quad \text{for} \quad \theta = 0;\ z = 0 \tag{3-91}$$

$$\psi = 1 \quad \text{for} \quad \theta = 2\pi;\ z = 0 \tag{3-92}$$

For all other horizontal levels, however, we must obtain values of the stream function at the points θ equal 0 and 2π numerically. Here one has to keep in mind that these two values of the angular coordinate correspond physically to the same point. For continuity reasons the difference between the values of the stream function in these angular positions must equal unity (the same integral flow across every horizontal surface). Yet, in spite of the discontinuity in the values of the stream function the derivative of the stream function with respect to the angle must be the same for θ equal 0 and 2π (the same value of the axial velocity component).

The equation obtained was solved for the distribution of resistances given by Eqs(3-86)-(3-88) with the simplification that $f_1 = 0$. This simplification has been done in order to obtain valid results independently of the mean velocity V_0 and it is justified so long as we remain in the region of asymptotically fast flows ($Re > 150$). Such solutions were obtained for several ratios of diameters d_c, d_a of particles composing the bed; namely 50/5.3, 40/7, 35/9.

In all cases investigated [36] the streamlines exhibited a wavy character in the direction of the flow as the fluid tended to react to the periodically changing structure of the bed. It is stressed though that the "waves" of the streamlines remain in the cylindrical surfaces determined by the radius for which the solutions are carried out.

The amplitude of the wavy pattern of the streamlines diminished with the decreasing ratio d_c/d_a. This, of course, could be expected as the diminishing ratio d_c/d_a means diminishing differences of the resistance of the zones occupied by different particles.

To characterize this flow rather more quantitatively, the following Fig.3-13 shows the dependence of the dimensionless mean axial velocity component, V_{z0}/V_0, Eq.(3-81), as a function of the angular coordinate θ. From the figure it may also be seen that the mean axial velocity component reflects the sinusoidal pattern of the structural inhomogeneity. The amplitude expectedly increases with the disparity of diameters of the two kinds of particles. On the other hand, it can also be seen that the deviations of the mean axial velocity from unity are not large and the mean standard deviation of V_{z0}, even for the largest size disparity reaches only 7.8% of the mean velocity.

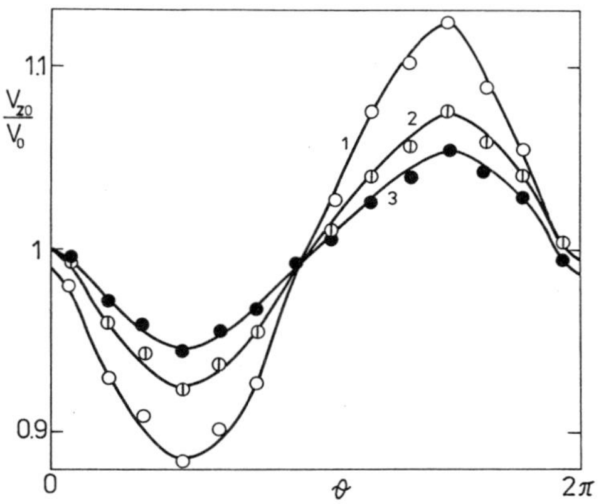

Fig.3-13
Dimensionless mean axial velocity component as a function of angular coordinate for a bed with angular distribution of resistance defined by Eqs (3-86) - (3-88). Empty circles - d_c/d_a = 50/5.3; halved circles d_c/d_a = 40/7; full circles d_c/d_a = 35/9 [mm/mm].

Table 3-2 summarizes the computed data regarding the energy losses in the structure of the bed investigated for the three particle diameter ratios. The second column in this table gives the dimensionless pressure drop per unit height of the bed ΔP^*, Eq.(3-82). These data suggest that the dimensionless pressure difference between the inlet and the outlet is almost independent of the disparity of particle size. The dimensional pressure drop, on the other hand, increases with the disparity of d_c and d_a. This is so because the dimensionless pressure drop in this case has to be multiplied by the mean resistance of the bed $(f_2)_0$, which is, of course, the highest for d_c/d_a = 50/5.3, as may be seen from the last column of the same table.

The dimensionless pressure drop per unit height of the bed, in addition, differs little from unity. From this it follows that the pressure losses in the investigated structures are almost the same as the losses in a uniform bed consisting of equal sized particles while this size is such that the integral mean resistance $(f_2)_0$, is the same.

A comparison of the second and the third column of the table points to the potential of diminishing the pressure losses through structural changes. The third column indicates the computed pressure losses for the case in which half of the bed (between angles 0 and π) is filled by particles d_c while the remaining half by particles d_a. Under such circumstances the flow is parallel and the pressure losses minimal and may be given Eq.(3-75).

Table 3-2

d_c/d_a [mm/mm]	Dimensionless pressure drop per unit height of bed ΔP^*	Theoretical minimum pressure drop per unit height of bed computed from Eq.(3-75) for given d_c, d_a	$(f_2)_0 = \frac{k}{2}\left(\frac{1}{d_c} + \frac{1}{d_a}\right)$
50/5.3	0.984 3	0.436 4	0.104 3
40/7	0.984 0	0.592 3	0.083 9
35/9	0.988 9	0.720 5	0.069 8

In conclusion one can say that layered beds with sinusoidally variable thickness of individual layers exhibit a relatively uniform axial flow while providing for a certain degree of mixing within cylindrical surfaces. This mixing occurs thanks to the "wavy" character of the streamlines. The calculations described [35,36] were all carried out for the ratio (R/r) equal to unity, that is for the peripheral region, but the picture would be at least qualitatively the same for other (R/r) ratios. Because of the relative uniformity of the flow the pressure savings are almost none irrespectively of the size ratio of the particles employed.

3.4.6. Potentials of structuralization of the bed for energy savings and associated flow maldistribution

It is clear now that potential energy savings achieved at the expense of the uniformity of the flow are not restricted to cases where the nonuniformity of the structure is such that the resulting flow is parallel.

In this paragraph we shall examine [72] structural nonuniformities retaining axial symmetry, which, however, do not ensure parallel flow, and will assess their potentials for energy savings and corresponding flow maldistribution. Let us consider a cylindrical bed sketched in Fig.3-14 of radius R charged with particles of two different diameters, say, d_a and d_b. The mode of charging these particles into the bed may be envisaged in such a way as if the bed were divided by a hypothetical cylindrical surface of radius R equal $0.7R$ coaxial with the bed. One kind of the particles is then first filled into the part near the axis of the bed up to the hypothetical cylindrical surface; the other particles are charged into the peripheral region between the wall and the hypothetical cylindrical surface. The height of the bed filled in this manner is $h/2$. As the next step the particles charged into the axial and peripheral region are switched. This procedure gives rise to a pattern of height h regularly alternating along the whole height of the bed. The regions of the bed occupied by the particles of different diameter are distinguished in Fig.3-14 by blank (smaller diameter particles) and black (larger diameter particles) regions.

We note that with the radius of the hypothetical separating cylindrical surface equalling $0.7R$ the volumes of regions occupied by the two types of particles in an arbitrary horizontal layer of height $h/2$ are nearly the same and if the limits are taken as shown in Fig.3-14 are exactly the same. We note further that the structure sketched is analogous to the structure designated as structure 4 in Fig.3-6. In contrast to the previous case, however, we shall now consider an infinitely deep bed.

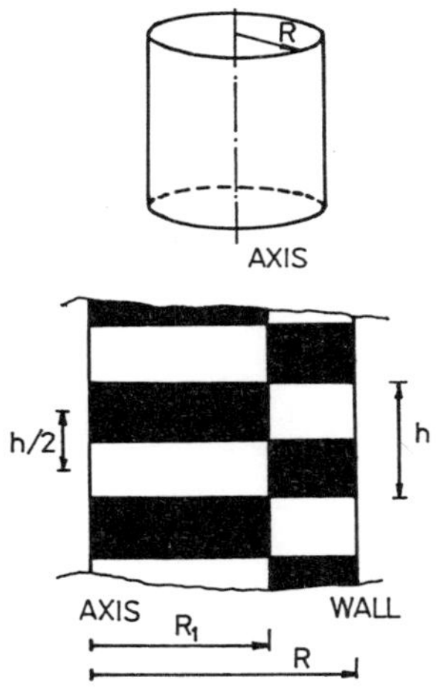

Fig.3-14
Sketch of the bed with axially symmetric distribution of resistance. Empty and full areas distinguish regions of different resistance.

Under such conditions it should suffice to solve the flow distribution equation for the section of the bed of height shown in Fig.3-14 utilizing the periodicity equation (the distribution of the flow on the top and bottom surface of the region must be the same). Even more effectively one can utilize the symmetry condition and solve the problem for the section of height $h/2$ only, provided its limits are taken as in Fig.3-14, making use of the fact that the top and bottom of this section are isobaric surfaces. This was actually the case in this study [72] and corresponding figures show the results in this section only.

It is now clear that the character of the flow and the pressure field in this section will depend on its relative height, that is on the magnitude of the simplex h/R. It may be anticipated that for low values of this simplex the structure of the nonuniformity alternates along the height of the bed so frequently that the gas (or fluid in general) is not given sufficient space to redistribute in response to the newly encountered nonuniformity profile. Consequently, the smaller the simplex h/R, the smaller the deviations of the flow from the uniform distribution.

Such a situation is depicted in Fig.3-15 showing the pattern of the streamlines for the relative height $h/R = 0.5$ while the ratio of diameters of the two types of particles composing the bed equals $d_a/d_b = 4:1$. The vertical and horizontal lines across the field of streamlines indicate the position of the hypothetical surface separating zones occupied by the larger and the smaller particles.

The corresponding pattern of isobars is shown in Fig.3-16. The isobar pattern exhibits substantial radial pressure gradient along the hypothetical cylindrical surface separating the axial and the wall region. This is readily explained by the fact that the flow for relative height of the periodic nonuniformity h/R is almost uniform and consequently the surface of the discontinuity of local resistances must give rise to large pressure gradients. In the same figure one can further observe the higher "density" of isobars in regions of higher local resistance, i.e. the lower left hand and the upper right hand corner of the figure.

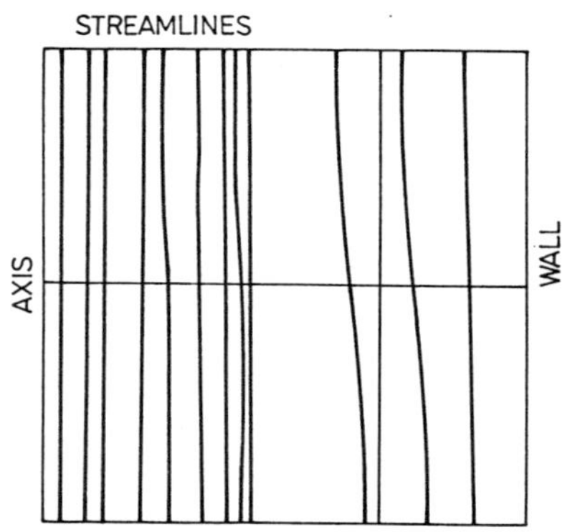

Fig.3-15
Streamline pattern for the structure in Fig.3-14 and $h/R = 0.5$. Thin lines indicate limits between zones of different resistance. $d_a/d_b = 4:1$.

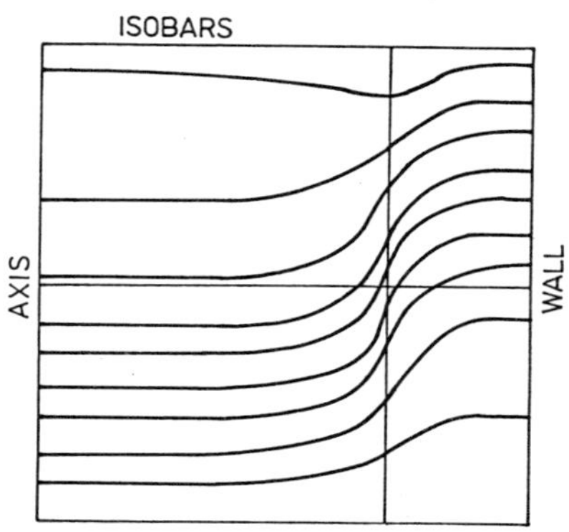

Fig.3-16
Isobar pattern for the structure in Fig.3-14 and $h/R = 0.5$. Thin lines indicate limits between zones of different resistance. $d_a/d_b = 4:1$.

A different picture of the situation is obtained by solving the same problem for $h/R = 8$. Fig.3-17 shows the streamline pattern for this case. The ratio of the smaller and the larger particles is the same as before, i.e. $d_a/d_b = 4:1$. The relative height of the periodic structural nonuniformity is now fairly large to allow the gas to adjust its flow distribution to the newly encountered resistance profile. As a result the flow near the top and the bottom of the section of the bed examined is nearly parallel although highly nonuniform. The part with appreciable cross flow is relatively short and compressed along the horizontal hypothetical separating surface.

In contrast, the corresponding isobars, shown for this case in Fig.3-18 are nearly horizontal lines suggesting spatially uniform energy dissipation. The fact that the gas was given sufficient room to adjust its distribution to the new profile caused almost all radial pressure gradients to vanish.

To demonstrate the effect of the cross flow, i.e. the role of the radial velocity component, Fig.3-19 shows the pattern of contours of constant radial velocity component for $h/R = 1$. These contours form closed curves around the intersect of the two hypothetical separating surfaces where maximum cross flow has been anticipated.

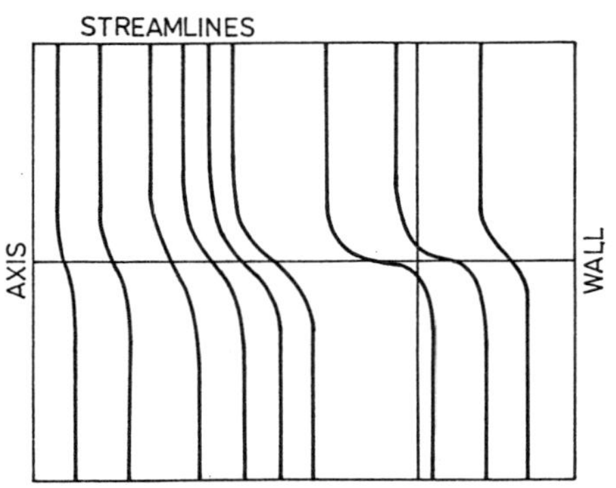

Fig.3-17
Streamline pattern for the structure in Fig.3-14 and $h/R = 8$. Thin lines indicate limits between zones of different resistance. $d_a/d_b = 4:1$.

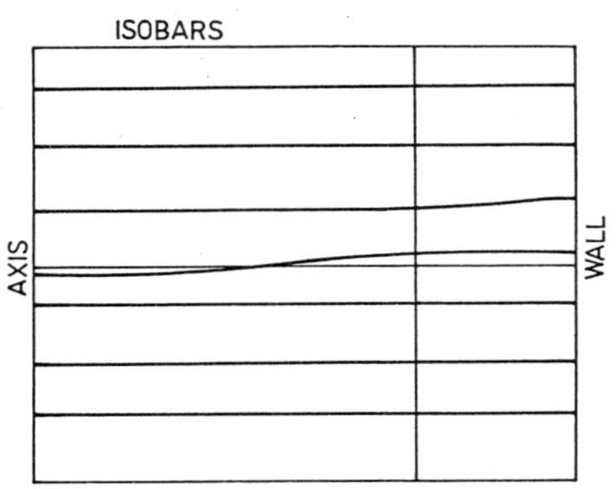

Fig.3-18
Isobar pattern for the structure in Fig.3-14 and $h/R = 0.5$. Thin lines indicate limits between zones of different resistance. $d_a/d_b = 4:1$.

The degree of maldistribution associated with the nonuniformities so far investigated is characterized by the criterion, s, defined in Eq.(3-80). This nondistributed characteristic of maldistribution is plotted in Fig.3-20 as a function of the relative height of the periodic structural nonuniformity h/R with d_a/d_b as a parameter.

For reasons apparent from the maps of streamlines previously shown it is obvious that the degree of maldistribution increases with the relative height of the periodic nonuniformity h/R and the increase is the more conspicuous the larger the disparity of particle diameters. With increasing h/R increases the fraction of the bed height where highly nonuniform although parallel flow prevails.

The next Fig.3-21 is a plot of the dimensionless pressure drop, Eq. (3-82), as a function of h/R again for various particle diameter ratios. In contrast to the flow maldistribution, increasing h/R ratio reduces the bed pressure drop making such beds more attractive from the energy input point of view.

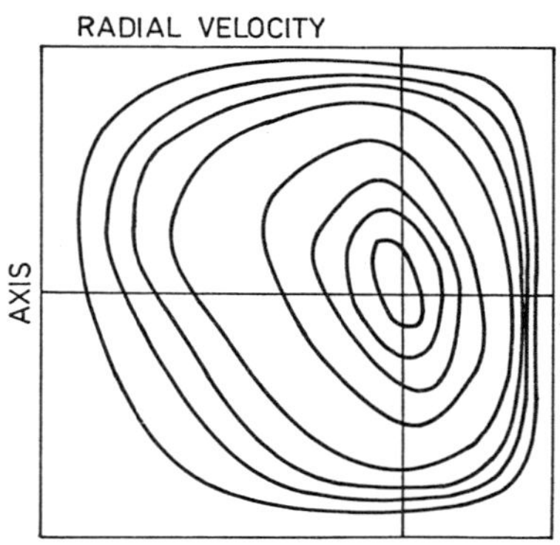

Fig.3-19
Contours of constant radial velocity component for the structure in Fig.3-14 and $h/R = 1$. Thin lines indicate limits between zones of different resistance. $d_a/d_b = 4:1$.

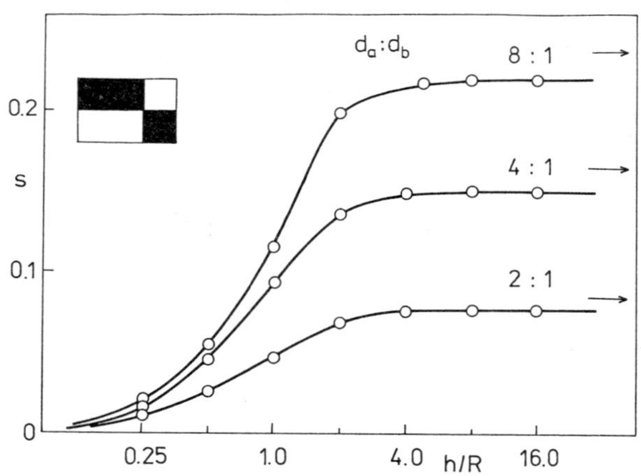

Fig.3-20
Criterion of flow nonuniformity for the structure shown in Fig.3-14 as a function of h/R and various particle disparities d_a/d_b .

However, for an efficient operation of the bed we usually require not only a low pressure drop but also a low degree of maldistribution. These two effects work against each other through the existence of the cross flow. Each particular case thus requires optimization between a tolerable degree of maldistribution and the energy costs of the operation.

These considerations can be carried even further by creating even more complex structures of structural nonuniformity. Another type is shown in Fig.3-22. The radii R_2 and R_3 are chosen so as to divide the bed cross section into sections of equal cross sectional area. Each horizontal section of height h or $h/2$, as shown in this figure, then contains the same volume of the two components of the charge distinguished by the black and blank regions.

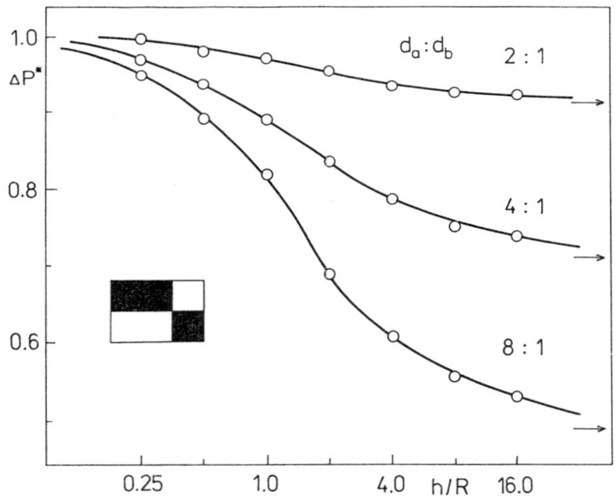

Fig.3-21
Dimensionless pressure drop for the structure shown in Fig.3-14 as a function of h/R and various particle disparities d_a/d_b .

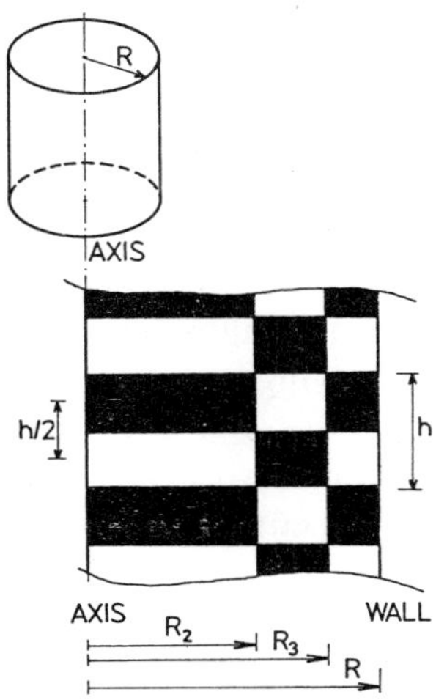

Fig.3-22
Sketch of the bed with axially symmetrical distribution of resistance. Empty and full areas distinguish regions of different resistance.

Computed patterns of the streamlines deviate less from the uniform flow than the patterns investigated previously for the same h/R and d_a/d_b ratios. The maldistribution was in this case confined to smaller scale. Corresponding patterns of isobars are more complicated exhibiting large radial pressure gradients along the cylindrical surfaces of radii R_2 and R_3 for small h/R ratios. Also, patterns of contours of constant radial velocity component exhibit typically two sets of closed contours, one corresponding to the negative, the other to the positive radial velocities. Both sets have their centers near the intersects of the hypothetical separating surfaces.

A comparison of pressure drops for the two structures of inhomogeneity examined in this paragraph for $d_a/d_b = 4:1$ is shown in Fig.3-23. For reasons associated with the flow maldistribution the savings of energy are even larger for the more complicated structure up to h/R equal to about 6. Ultimately, however, the savings for very high h/R are smaller as the more complex structure restricts large scale flow nonuniformity.

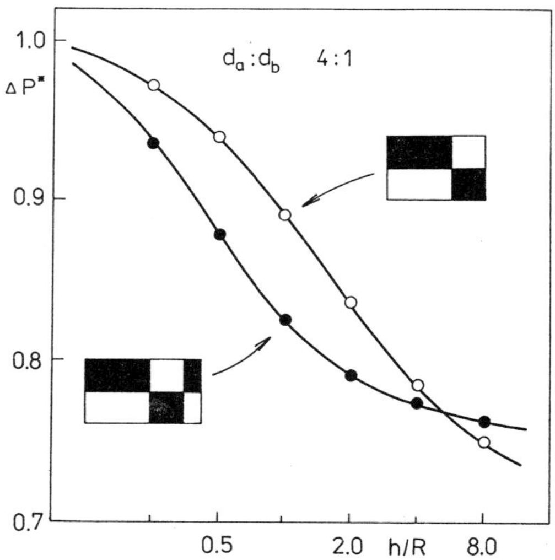

Fig.3-23
Dimensionless pressure drop as a function of h/R for two configurations of axially symmetric resistance distribution.

The computed results have shown that structurally nonuniform beds can be operated at lower pressure drops even under generally nonparallel flow. The flow uniformity and energy losses work against each other. The maximum degree of maldistribution that is tolerable depends on the type of underlying technological operation carried out in the bed.

In beds where chemical reactions take place the requirements on the flow uniformity can often be very strict. Nevertheless, the requirement may often be such as to keep the residence time the same for all elements of the fluid in the bed. Such requirements, of course, are relatively well met by the structures investigated in this paragraph because the axial velocity component averaged along the height of the bed varies relatively little with the radius. It does often vary significantly, of course, along the vertical coordinate. This variation, however, need not necessarily adversely affect the chemistry of the process within the bed.

In other types of packed bed operations the two types of particles differ not only in size but are also chemically different. This is often the case in metallurgy or in gas-solid reaction systems. In these cases one has to keep in mind that appreciable cross flow exists only in the neighbourhood of separating surfaces. The fluid then travels alternately through regions of chemically different particles without appreciable cross flow. The relative height h/R has then to be chosen in each particular case individually keeping the effects on the chemistry of the process in mind.

Throughout this paragraph no provisions have been made for variations of the void fraction along the surfaces of contact between regions of different particles and associated pressure losses. These effects will be dealt with in the next paragraph.

3.4.7. The flow across "interfaces" between layers of particles of different size

The question of the flow and pressure losses in beds with a complicated structure, such as those investigated in the previous paragraphs, cannot be answered sufficently accurately unless also structural changes at the surfaces of contact between zones of the bed occupied by particles of different diameter are taken into account. As has been shown in chapter 2 these "interfacial" regions exhibit diminished void fraction and, accordingly, one has to expect additional pressure losses in comparison with the flow across an ideal interface where particles do not mingle.

Direct measurements of "interfacial excess" pressure drop have been carried out by Szekely and Propster [51] on a 0.97 m in diameter bed with technical particles of coke, ore sinter, etc. The interfacial excess pressure drops were evaluated from extrapolations of the pressure versus distance curves from the pure component regions onto the interfacial plane. The excess pressure drop is thus attributed to an idealized plane separating the two adjacent layers composed of different materials. This is certainly a good means of circumventing the problem of determining the extent of the interfacial region where the mixing of the two kinds of particles takes place.

Szekely and Propster [51] found the excess pressure drop to be proportional to the square of gas linear velocity which implies that the inertial term in the Ergun equation in the interfacial region dominates. The excess pressure drop from their data can then be computed from the inertial term of the Ergun equation in which the extent of the "interfacial" region is taken the diameter of the particles forming the bottom layer (d_b) while the diameter and the shape factor of particles within the "interfacial" region are taken as arithmetic averages of the particles forming the two adjacent layers. The void fraction of the interfacial region is taken to be the void fraction of the mixture of the two components. Thus

$$\Delta P_{\text{interf}}^{\text{excess}} = 1.75 \frac{(1 - \varepsilon_{\text{interf}})}{\varepsilon_{\text{interf}}^3} \frac{2d_b}{(d_t + d_b)} \varrho V_0^2 \tag{3-93}$$

where d_t, d_b also incorporate appropriate shape factors.

Staněk and Eckert [21,28,43] based the calculation of the excess pressure drop due to the flow across the region of contact of particles of different diameter on the assumption that the increased pressure losses appear mainly due to the locally decreased void fraction. Further it was assumed that such pressure losses obey, in principle, the scalar form of the Ergun equation. The excess pressure drop may then be obtained from

$$\Delta P_{\text{interf}}^{\text{excess}} = 150\mu E_1 V/d_e + 1.75\varrho E_2 V^2 \tag{3-94}$$

where the dimensionless coefficients E_1, E_2 are given by the following integrals

$$E_1 = \frac{1}{d_e} \int_{h1}^{h2} \left[\frac{(1 - \varepsilon)^2}{\varepsilon^3} - \frac{(1 - \varepsilon_{\text{id}})^2}{\varepsilon^3} \right] \mathrm{d}z \tag{3-95}$$

$$E_2 = \frac{1}{d_e} \int_{h1}^{h2} \left[\frac{(1 - \varepsilon)}{\varepsilon^3} - \frac{(1 - \varepsilon_{\text{id}})}{\varepsilon^3} \right] \mathrm{d}z \tag{3-96}$$

d_e here designates an equivalent particle diameter in the "interfacial" region to be specified later.

The integration limits h_1, h_2 are taken so as to encompass the whole "interfacial" region where variations of the void fraction occur. Subscript "id" designates the area void fraction where no mingling of particles occurs.

The vertical coordinate of the surface separating the two layers falls naturally between the limits h_1, h_2. The practical advantage of the above choice of the definition of the coefficients E_1, E_2 is that the expression for $\Delta P_{\text{interf}}^{\text{excess}}$ is virtually insensitive to the accuracy of the determination of the extent of the "interfacial" region and therefore to the choice of the limits h_1, h_2.

A large number of experiments [21] were carried out with glass spheres 3,4, 5,10,15 and 20 mm in diameter and processing of the void fraction profiles between layers of these particles arranged always in such a way that the smaller particles formed the top layer were smaller than the spheres forming the bottom layer ($d_t/d_b < 1$) yielded values of the coefficients E_1, E_2.

Based on geometrical argument the results on double layers of spheres are expected to be the same for those runs where the ratio d_t/d_b remains the same. Other differences than those stemming from the ratio of particle diameters could perhaps appear due to the different way of preparing the double layer and, in case of small vessels, through the presence of containing walls.

The coefficients E_1, E_2 of Eq.(3-94) were correlated for $d_t/d_b < 1$ with the following result

$$E_1 = \left(\frac{d_t}{d_b}\right)^{-2.649} - 1 \tag{3-97}$$

$$E_2 = \left(\frac{d_t}{d_b}\right)^{-2.858} - 1 \tag{3-98}$$

The standard deviations of the exponents in the above correlations amounted to 0.0877 and 0.1074 respectively.

The results were [21] somewhat different if the larger particles were charged on top of the smaller ones, i.e. if $d_t/d_b > 1$. Analogous correlation for the coefficients E_1, E_2 for $d_t/d_b > 1$ yielded following results

$$E_1 = \left(\frac{d_t}{d_b}\right)^{-2.253} - 1 \tag{3-99}$$

$$E_2 = \left(\frac{d_t}{d_b}\right)^{2.027} - 1 \tag{3-100}$$

The validity of these correlations was further restricted to $(d_t/d_b) > 2.42$. Below this limit, the coefficients are taken to be zero.As noted in chapter 2, for small disparity of particle size, particle mingling could not occur and the found values of E_1, E_2 were zero within experimental error. The limiting value of the disparity ratio was evaluated from the expression for the mean pore diameter expressed in terms of sphere diameter as $0.62\ \varepsilon/(1 - \varepsilon)$ found by Zagrafskaia [78].

Another study [28,77] investigated the excess "interfacial" pressure drop for double layers of technical particles of coke (two fractions) and agglomerate (seven fractions). These fractions were relatively narrow and certainly contained no fines or oversized particles. The mean size of particles in each fraction was determined by an image analyser from the first moment of the projected areas of all particles.

Using an analogous argument regarding the dependence of the excess pressure drop on the geometry of the double layer the results of the coefficients E_1, E_2 of Eq.(3-94) were correlated for $(d_t/d_b) < 1$ with the following result

$$E_1 = \left(\frac{d_t}{d_b}\right)^{-1.859} - 1 \tag{3-101}$$

$$E_2 = \left(\frac{d_t}{d_b}\right)^{-1.639} - 1 \tag{3-102}$$

with the standard deviation of the exponents 0.1971 and 0.2226 respectively.

The correlation for the case with the bottom layer formed by smaller particles, i.e. $(d_t/\mathrm{d}_b) > 1$, then reads

$$E_1 = 0.312\,1 \left(\frac{d_t}{d_b}\right)^{2.647} - 1 \tag{3-103}$$

$$E_2 = 0.182\,5 \left(\frac{d_t}{d_b}\right)^{2.791} - 1 \tag{3-104}$$

The coefficients in front of the particle diameter ratio indicate that below a certain limit the coefficients E_1, E_2 would be negative. These limits are $(d_t/d_b) = 1.552$ and 1.839 for E_1 and E_2 respectively. This again is the consequence of the fact that for small particle disparities no mingling of particles occurs and here we assign the coefficients zero values.

The equivalent particle diameter for Eq.(3-94) was in all last eight correlations computed as

$$\frac{1}{d_e} = \left(\frac{1}{d_t} + \frac{1}{d_b}\right) \tag{3-105}$$

A comparison of Szekely's correlation, Eq.(3-93), from direct excess pressure drop measurement, with the correlation (3- 94), based on the coefficients E_1, E_2, shows a good agreement between results based on measurements with spheres, Eqs(3-97) and (3-98). The explanation rests probably on the fact that narrow fractions of technical particles, Eqs(3-101) and (3-102), displayed weaker dependence on the disparity of the particles forming the two layers. This is probably the consequence of the rough surface of the particles restricting their easy penetration and mingling between layers. Stronger dependence on the disparity ratio of Szekely's results is mainly due to the presence of the fines. An exponent over the particle diameter ratio estimated from Szekely's data amounts to -2.424, i.e. a value close to that found for spheres, Eq.(3-98).

A general conclusion is that the excess "interfacial" pressure drop is especially large when smaller particles are charged on top of very large ones and the interfacial region may then contribute significantly to overall pressure losses.

3.4.8. The flow through structuralized beds in the presence of "interfacial" resistances

The existence of excess resistances in regions of contact of zones of particles of different size will, however, cause not only an increase of the net energy requirements for the flow but will also modify the field of velocities and isobars.

Let us consider again structures A,B,C sketched in Fig.3-12 and investigated, as far as their flow distribution and pressure drops are concerned, provided that there are no additional losses associated with the crossing of the gas over the "interface" between regions occupied by particles of different size.

Let us introduce cumulative resistance functions in the case of axially symmetric problems for the directions of the coordinates of radius r and the axial coordinate z, in the following manner

$$F_{1r} = \int_0^r f_1 \mathrm{d}x \tag{3-106}$$

$$F_{2r} = \int_0^r f_2 \mathrm{d}x \tag{3-107}$$

$$F_{1z} = \int_0^z f_1 \mathrm{d}x \tag{3-108}$$

$$F_{2z} = \int_0^z f_2 \mathrm{d}x \tag{3-109}$$

The above shown integration can be easily carried out analytically for the structures A,B,C and , in fact, for all structures investigated in all previous paragraphs. Differentiation of these cummulative resistances with respect to the appropriate coordinate yields then local values of the resistance coefficients f_1, f_2. This approach is particularly convenient when finite difference methods are applied and calculations with gradually refined computational grids are performed. Refined computational grids provide ever more accurate description of the step change of the resistance coefficients on transition from one layer to another.

In the case now investigated the "interfacial" region is the source of additional resistance to the flow which can be conveniently incorporated into the cumulative resistance function.

If we consider [42] that the extent of the "interfacial" region is small compared to the size of individual layers of the structured bed the excess pressure drop, say in the direction of the coordinate z, can be written as one component of the vectorial form of Eq.(3-94) in the following form

$$\Delta P = V_z \int_{h1}^{h2} (f_1 - f_{1id})\, dx + V V_z \int_{h1}^{h2} (f_2 - f_{2id})\, dx \tag{3-110}$$

where "id" subscripted resistance coefficients f_1, f_2 correspond again to an ideal bed free of the excess pressure losses at the interface.

Using the definitions Eq.(3-95) (3-96) of the coefficients E_1, E_2 one can easily modify the definitions of the cumulative resistances to obtain

$$F_{1r} = \int_0^r f_1 dx + 150\mu \frac{E_1}{d_e} \tag{3-111}$$

$$F_{2r} = \int_0^r dx + 1.75\varrho E_2 \tag{3-112}$$

$$F_{1z} = \int_0^z f_1 dx + 150\mu \frac{E_1}{d_e} \tag{3-113}$$

$$F_{2z} = \int_0^z f_2\, dx + 1.75\varrho E_2 \tag{3-114}$$

In this approach [42] we have tacitly assumed (see Eq.(3-110) that changes dominating the "interfacial" region are those of the resistance, not the velocity vector.

The correlations used for the evaluation of the coefficients E_1, E_2 have to respect which particles (smaller or larger) are on "top" judged with respect to the direction of gravity.

In addition, the second term in the four last definitions has to be added repeatedly for each interface existing between the integration limits of the integral of the first term, while the value of the coefficient E_1 or E_2 will probably be different for each interface crossed.

The mathematical representation of the cumulative resistances reduces the interfacial regions to mere surfaces, which is acceptable as the physical dimensions of the "interfacial" region will be mostly small compared to the physical dimensions of individual layers.

Modified cumulative resistances in Eqs.(3-111)-(3-114) can be again evaluated in most cases analytically [42]. Numerical differentiation of the cumulative resistances, computed for individual points of the computational grid from these analytical expressions, provides then local resistance coefficients. Local resistance coefficients computed in this manner exhibit narrow peaks in regions of contact of particles of different diameter, while the width of these peaks steadily narrows with the refinement of the computational grid.

Differentiating the cumulative resistances twice provides the derivatives of local resistance coefficients required for the solution of the flow equation (3-11) with the same favourable property regarding the refinement of the numerical grid.

Such a procedure has been employed [43] also in case of the structures A,B,C sketched in Fig.3-12. The coefficients E_2 were evaluated from the correlations (3-98) and (3-100) as we have investigated the case of large superficial gas velocities. The results in the dimensionless form are then independent of the velocity V_0.

The diameters of particles of the two components occupying regions 1 and 2 in Fig.3-12 were taken to be 51.4 mm and 12.4 mm. The d_t/d_b ratio thus amounted to 4.1 or 0.24 depending on which of the particles were on "top".

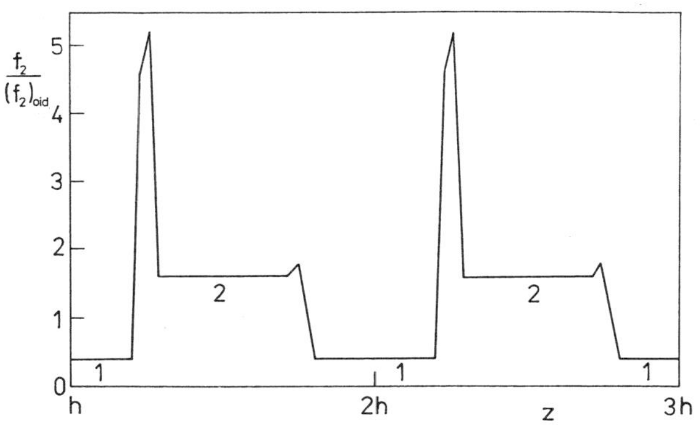

Fig.3-24
Dimensionless local resistance coefficient including "interfacial" resistances as a function of axial coordinate for radius $r = 0.5R$ and the structure B from Fig.3-12.

Fig.3-24 shows the course of the local resistance f_2 for the structure B of Fig.3-12 along the vertical coordinate on the radius $r = 0.5R$. The component "1" in this case are the larger particles and therefore the local resistance is low. The first peak from the left represents the large resistance associated with the "interface" when the smaller particles are on top of the larger ones. The peak is followed by a plateau corresponding to the part of the bed occupied by the larger particles. This plateau is terminated by a small peak associated with the "interfacial" resistance when the larger particles are on top of the smaller ones. The whole sequence then repeats.

The plotted local resistance is scaled by the mean resistance $(f_2)_{0id}$ of the bed designating the mean resistance defined in Eq.(3-76) which does not include the interfacial resistances ("ideal" layer). The mean resistance of an ideal layer in this form is equal to unity and Fig.3-24 clearly indicates the local increase at the interface of the layers of different particles.

Table 3-3

Run No.	Structure	$\frac{h}{R}$	d_1 [mm]	d_2 [mm]	$\frac{(f_2)_0}{(f_2)_{0id}}$
1	2	3	4	5	6
1	A	0.36	51.45	12.42	1.265
2	A	0.36	12.42	51.45	1.267
3	B	0.36	51.45	12.42	1.266
4	B	0.36	12.42	51.45	1.266
5	C	0.36	51.45	12.42	1.267
6	C	0.36	12.42	51.45	1.265

Table 3-3 continuation

$\frac{(f_2^{-0.5})_0^{-2}}{(f_2)_0}$	$\frac{(f_2^{-0.5})_0^{-2}}{(f_2)_{0id}}$	s	$\frac{V_{z0}(r = R)}{V_{z0}(r = 0)}$	ΔP^*
7	8	9	10	11
0.774	0.979	0.096 4	1.4	1.245
0.798	1.011	0.126 3	0.64	1.190
0.785	0.994	0.015 7	0.98	1.268
0.782	0.990	0.015 5	0.98	1.225
0.783	0.992	0.140 0	0.53	1.205
0.784	0.992	0.116 0	1.54	1.194

Numerical solution of the flow in structures A,B,C, including the interfacial resistances, was carried out for six cases summarized in Table 3-3. In all cases the height of a single charge layer was equal $0.36R$. The increase of the mean resistance of the layer due to the existence of interfacial resistances amounts to about 26% as may be seen from the sixth column (these data should be the same for all six runs; differences stem from the numerical procedure of evaluating the integral average).

The streamlines calculated for all runs in Table 3-3 display largely vertical direction with slightly wavy character. Fig.3-25 indicates contours of constant component of velocity V for the structure A and run No.1. Since in this case the larger particles were charged first (see the fourth and the fifth column of this table) the region near the wall has a greater portion of the larger particles and therefore smaller resistance. Correspondingly, the contours of constant V_z, representing higher values appear in the right hand side of the figure. It is also interesting to note that the contours of V_z display high curvature in regions where they cross the "interface" between layers of different particles. These interfaces are shown in the figure by broken lines.

Fig.3-26 presents again contours of constant V_z computed this time for the structure B and run No.3 in Table 3-3. All contours in this figure are contours of the dimensionless axial velocity V_z/V_0 equal 0.9. Regions denoted by the plus sign are regions where the axial velocity component is larger than 0.9; regions marked by the minus sign are regions where the axial velocity component does not reach 0.9. Since the contours of constant V_z are approximately parallel with the borders between regions charged by particles of different diameter (shown by broken lines) the mean axial velocity component is almost independent of the radial coordinate and the mean flow in the bed is largely uniform.

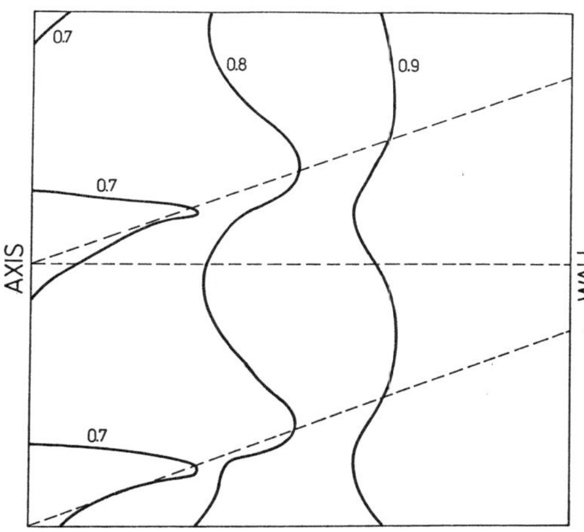

Fig.3-25
Contours of constant dimensionless axial velocity component for the structure A from Fig.3-12 (run No 1 in Table 3-3). Figures on curves indicate V_z/V_0; broken lines indicate limits between zones occupied by different particles.

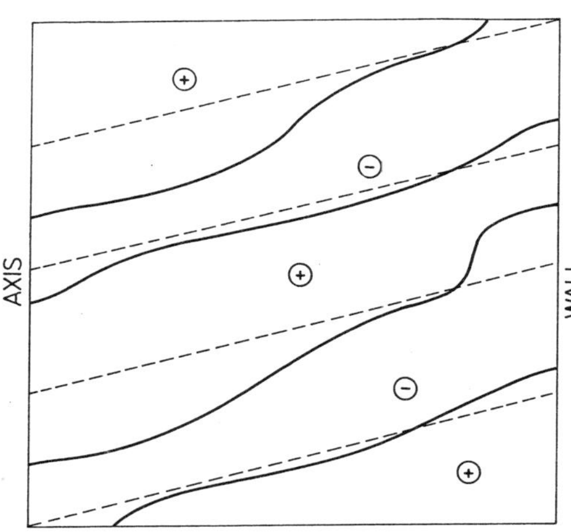

Fig.3-26
Contours of constant dimensionless axial velocity component for the structure B from Fig.3-12 (run No 3 in Table 3-3). All contours are contours of $V_z/V_0 = 0.9$; plus/minus sign indicates region of higher/lower velocity.

As already mentioned the increase of the mean resistance of the bed due to the "interfacial" resistances amounts to about 26%. This increase, however, depends on the frequency of alternating zones of particles of different diameter, i.e. on the dimensionless height of the layer of a single charge, taken in all runs in Table 3-3 to be 0.36. If the height of the layer of a single charge is doubled $(0.72R)$ the relative increase of the mean resistance $(f_2)_0$ amounts to only 9.5%. On the other hand, for more frequent changes of the structure the increase of the resistance would be higher.

The seventh and the eighth column of Table 3-3 indicate quantities representing dimensionless pressure drop if the flow on each horizontal level had been redistributed according to the local distribution of the resistances f_2. In the seventh column these dimensionless pressure losses are expressed (see the definition in Eq.(3-82)) as fractions of the losses of the bed with a uniform distribution of resistances equalling $(f_2)_0$, while in the eighth column these losses are scaled by the losses of the uniform bed with the mean resistance $(f_2)_{0id}$.

Since the values in the eighth column are all close to unity it is seen that the theoretical minimum of pressure losses achievable by structuralization of the bed and the nonuniform flow through these structure, is almost completely offset by increased losses incurred by the flow across the "interfacial" region between particles of different diameter. The investigated dimensionless height of the layer of a single charge 0.36 can be taken therefore as a critical value.

The computed dimensionless pressure losses are given in the last, eleventh column of Table 3-3. (The scaling factor in this case is $(f_2)_{0id}$ and these data must therefore be confronted with the eighth column of the table. Energy savings given rise to by the nonuniform flow follow from comparison of columns 11 and 6.

The ninth column of Table 3-3 indicates the nonuniformity criterion, defined in Eq.(3-80). Expectedly, as in the ideal case free of the excess pressure losses, the uniformity of the flow is good for structures B but worse for structures A and C, characterized by the radial nonuniformity of the resistances.

The ratio of the mean axial velocity component, V_{z0}, near the wall and in the axis is shown in the tenth column of Table 3-3. Expectedly, structures with the larger particles mostly near the wall (structure A run No.1 and structure C run No.6) exhibit values of this ratio higher than unity suggesting (as in Fig.3- 25) predominantly peripheral flow. On the other hand, structure depositing smaller particles largely near the wall (structure A run No.2 and structure C run no.5) give rise to flows concentrated around the axis.

It is interesting to note that upon correcting the result of the pressure drop calculations to the same particle diameter ratio (d_1/d_2) one finds, from comparison of the pressure drops for the case with and without "interfacial" resistances and otherwise the same conditions, that the existence of excess interfacial resistances nearly always further impairs the uniformity of the flow.

It is also interesting to note that Szekely and Propster [51] found that for conventional layer-charged blast furnace burdens, resembling the structure B in Fig.3-12 and fixed ratio of the height of the layer of a single charge, offer the least resistance to flow. More frequent changes of the structure naturally increase the mean resistance of the bed and may eventually exceed the pressure losses of a uniform layer where both components are completely mixed in the same proportion. From this finding one can conclude that the structure of the mixture of particles in the "interfacial" region represents the most compact structure possible, thus maximizing the associated pressure losses.

A general rule for charging practices therefore is that the effect of "interfacial" resistances is especially profound when smaller particles are placed on top of the larger ones and placing very small particles on top of the large ones should therefore be avoided as much as possible.

Schultz and Abel [157,158] actually used "specific" pressure loss, i.e. local pressure loss per unit height, to detect the extent of the "interfacial" region where these losses are significantly higher.

In reality, of course, the conditions in the burden, or generally structuralized beds are more complicated than shown by the model calculations in the preceding paragraphs. Kajiwara et al. [112] investigated these conditions on a cold experimental model of a blast furnaces and found that a mixed layer of particles forms during charging and presented a correlation for the increase of the thickness of the coke layer due to the appearance of the mixed zone. Secondly, they detected a decrease of the angle of deposit of the coke layer due to the effect of gas flow on the burden distribution. Thirdly, they found that there is a velocity distribution of the burden descent and that the descent velocity near the wall is usually smaller . Of these factors the most significant effect on burden distribution was found [112, 151,152,157,158] to be the formation of the mixed layer. Mixing of coke and sinter layers takes place at high gas rates also due to fluidization of the burden, see e.g. Kuwano et al. [159].

In an earlier paper Narita et al. [113] ascribed the change of the burden distribution to the fluidization of the coke layer in the central zone where a central coke column forms. In accordance with the analysis of the flow distribution in the previous paragraphs the appearance of the coke column within the granular zone of the furnace decreased the corresponding pressure drop in this zone by concentration of the gas flow to this region due to the increased nonuniformity of the flow.

An additional complication for the modelling of the real blast furnace is the existence of a low permeability cohesive zone. The lowered permeability is due to the softening of the burden material. Position of the cohesive zone is not horizontal, which thus contributes to the nonuniformity of the flow of gas both below and above the cohesive zone. Models to account for this effect also exist in the literature. See e.g. works of Hoshide et al. [160] and Gudenau et al. [161,162,163].

List of symbols to chapter 3

a	exponent in the viscosity function, Eq.(3-58)
b_1, b_2	resistance coefficients for a compressible fluid defined in Eqs (3-48), (3-49)
c_1, c_2, c_r	resistance coefficients for an ideal gas defined in Eqs (3-60), (3-61), (3-62)
d, d_a, d_b, d_c, d_t, d_1, d_2	particle diameter
d_e	equivalent particle diameter
d_0	mean characteristic particle diameter
$\mathbf{e}_r$, $\mathbf{e}_z$, $\mathbf{e}_\theta$	unit vectors in directions of cylindrical coordinates
E, E_v	rate of energy dissipation per unit length and per unit volume of bed
E_1, E_2	resistance coefficients for the flow across regions of contact of different particles defined in Eqs (3-95), (3-96)
f_1, f_2	resistance coefficients defined in Eqs (3-2), (3-3)
$(f_2)_0$	integral mean coefficient f_2 defined in Eq.(3-76)
$(f_2^{-0.5})_0$	integral mean of $f_2^{-0.5}$ defined in Eq.(3-35)
$\mathbf{G}$, G, G_z	superficial mass velocity vector, its magnitude and axial component
h	bed depth
k	constant relating $(f_2)_0$ to d_0 defined in Eq.(3-83)
M_g	molecular mass of gas
P, P_0	pressure and reference value
ΔP, ΔP^*	pressure difference dimensional and dimensionless defined in Eq.(3-82)
$\Delta P_{inter}^{excess}$	excess pressure drop due to flow across "interface"
r	dimensional coordinate of radius
R	radius of bed
R_g	versatile gas constant
Re	Reynolds number defined on V_0 and bed specific surface
Re_p	Reynolds number defined on V_0 and particle diameter
s	criterion of flow nonuniformity defined in Eq.(3-80)
t	time
T, T_0	absolute temperature and its reference value
U	quantity defined in Eq.(3-72)
$\mathbf{V}$, V, V_r, V_z, V_0	vector of superficial velocity, its magnitude and components in directions of cylindrical coordinates
V_0	superficial mean velocity
V_{z0}	velocity V_z averaged along the bed depth defined in Eq.(3-81)

X	volume fraction of bed occupied by coke particles
X_s	volume fraction of smaller particles in mixture
z	dimensional axial coordinate
μ, μ_0	viscosity and its reference value
ϱ, ϱ_0	density and its reference value at P_0
ψ	stream function
ε, ε_m	void fraction and void fraction of a mixture of particles
ε_b, ε_w	void fraction in bulk and wall region
θ	angular coordinate
grad, rot, div	vectorial operators of differentiation operating in dyadic, vector and scalar product

Additional symbols defined in text.

4. Velocity field under the trickling flow of liquid in a fixed bed

In the previous chapter we have examined the consequences of various types of structural inhomogeneities for the flow of a fluid in the interstices of the bed when the fluid completely filled the voids of the bed.

However, both theoretically and practically important and interesting are the situations in which the liquid trickles down the bed under the effect of gravity without filling entirely the voids in the bed. The remaining part is occupied by gas, which may be either stagnant or flow co- or counter-currently with the liquid. The gas may or need not take part in the process taking place within the bed.

Examples of equipment with this type of the flow of liquid include packed distillation and absorption columns, irrigated trickle bed reactors, bioreactors with immobilized enzymes, reactors for isotope exchange, power plant water cooling towers. Also the ironmaking blast furnaces in their lower part are essentially reactors of this type. The bed here consists mainly of coke particles with the liquid products of the process (molten slag and iron) trickling down with counter-currently streaming reducing gas.

For an accurate description of the transport phenomena and/or chemical reactions in such beds one needs information about the distribution of the flow of liquid in the bed.

Nonuniform distribution of the flow of liquid in the bed can be divided roughly into two categories: Small scale and large scale maldistribution. Of small scale effects one speaks at the level of particle size and associated maldistribution is connected with the fact that the liquid is not uniformly spread over the surface of the particle but often forms rivulets which frequently do not touch all particles. A phenomenon that counters this drawback is radial mixing.

Large scale maldistribution effects are associated with the inhomogeneity of the structure of the bed in the wall region where there is increased local void fraction, void fraction oscillations and vastly different radii of curvature of the surface of the particles and the confining walls. This in turn results in the preferential flow of liquid toward the wall and formation of the wall flow there. An additional cause of large scale liquid flow nonuniformities is the nonuniformity of distribution of liquid supplied at the top by the liquid distributor.

Various models have been developed in the literature to predict the distribution of liquid flow over the column cross section at various depths of the bed. All these models can be divided essentially into two categories:

Diffusion models. These models view the motion of the liquid through the packed bed as random motion of individual liquid elements forming a continuous medium. The motion of the liquid thus resembles diffusion of a solute in a solvent. The properties of the fixed bed and its structure are then accounted only in the "diffusion" coefficients characterizing the randomness of the motion of liquid. An entirely opposite approach is taken in:

Percolation principle models. Here the modelling concentrates on the paths of the fluid which are determined by the bed itself. The formation of the model paths is also in a sense "random" and the passage of the liquid through the structure of these paths is called the percolation process.

Although the percolation models are very promising for future progress in the modelling of liquid flow in packed beds this chapter largely concentrates on the diffusion models that have been elaborated to a greater depth to date. They also fall into the category of pseudohomogeneous models consistently pursued in this book.

Diffusion models attack the problem via solutions of differential equations of the diffusion type; hence their finite difference form encompasses also existing "cell" models, see e.g. Jameson [177].

4.1. THE PERCOLATION MODEL CONCEPT OF LIQUID FLOW IN A PACKED BED

The percolation theory was applied to the problem of packed beds for the first time by Crine, Marchot and L'Homme [191,192,197,198,211]. The bed in this case was represented by a lattice of sites and bonds.

The sites may be envisaged as the points of contact between individual particles of the packing. The liquid is transported from one site to another via a bond. The bonds thus may be envisaged as the surface of the packing particles.

The lattice is usually selected in the form of a "cubic body centered" lattice, although a number of other alternatives are available. For the "cubic body centered" lattice the site appears in the center of a cube with the bonds, transporting the liquid away from the site, being the diagonal bars connecting the site with the four bottom corners of the cube.

The first step of the modelling procedure is the generation of the medium transporting the liquid (the bed). This is achieved by assigning randomly logical values zero or unity to each bond thus setting up a Boolean matrix whose first three indices determine the spatial position of the site within the lattice, while the fourth index identifies the bond emanating from the site. In case of the cubic centered lattice the fourth index assumes values between 1 and 4.

The logical values zero or unity indicate whether the given bond is active for the transport of liquid (1) or inactive (0). The assignment is done in such a way that a value of the "permeability of the bed" is selected between zero and unity and a random number from the same range is generated for each bond. If the random number is less or equal to the selected permeability of the bed the particular bond is taken to be active and the corresponding element of the Boolean matrix is assigned a unity value. In the opposite case the bond is taken to be inactive and the corresponding element is assigned a zero value.

If a very low value of the permeability of the bed is selected the resulting matrix of active bonds is very sparse and the active sites form only isolated clusters of interconnected bonds. Above a certain threshold value of permeability individual clusters of active bonds cross the whole lattice in the direction of the flow, i.e. from the top to the bottom of the bed, creating thus a trajectory which can transport liquid through the bed and makes the bed actually permeable.

In case of the flow of liquid in a packed bed we are, of course, interested only in these very trajectories contributing to the permeability of the bed. As a next step we therefore proceed from the bottom layer of the lattice and eliminate all bonds that lead to the sides of the lattice (which in computer simulation cannot be taken as infinite in the direction transverse to the direction of the flow). In addition, we also eliminate the bonds leading to dead ends within the lattice.

Finally, random walks are simulated repeatedly from all active sites at the top level of the lattice while counting and marking all active bonds encountered. The procedure is repeated until the mean number of non-marked bonds encountered in one step is less than a certain level.

These results are then processed to yield the following quantities (see Crine et. al [191]):

The *irrigation rate*, taken to be the number of used active bonds expressed as a fraction of the total number of bonds in the lattice.

The *overlapping rate* of the trajectories, taken to be the number of bonds used more than once as a fraction of the total number of bonds used.

The *isotropy rate,* taken to be the fractional number of flow trajectories which are of infinite width (in the finite-size computer simulations the trajectories laterally bonded from wall to wall).

The simulation of the bed as outlined above implies certain assumptions: The randomly packed bed is assumed to be isotropic. The random choice of the bond taken to be active means that there is no interaction between the bonds.

Crine et. al [191] simulations showed the threshold permeability to be 0.38. Below this limit both the irrigation rate and the overlapping rate were zero; the bed was essentially impermeable. Above this limit both quantities steadily increased.

The isotropy rate was practically zero up to the bed permeability of about 0.7. Above this limit it sharply increased indicating transition from an anisotropic flow, characterized by finite width trajectories and hence more or less individual liquid paths, to infinite width clusters and corresponding isotropic flow.

Ahtchl and Pedersen [212] formulated what they call a Very Large Lattice Model of liquid mixing in trickle beds based also on the percolation concept. Although the algorithm differs relatively little, utilizing, in addition, the least congestion concept, their results differ from those of Crine et. al. The least congestion path concept means that the random choice of the bond to be followed by liquid from a certain site is made only from the bonds which had been used least. This leads naturally to minimizing the local flow densities.

The results of Ahtchl and Pedersen [212] indicated a steady increase of the irrigation rate from a threshold value of about 0.17. In contrast, however, the overlapping rate exhibited local minimum and maximum between the permeabilities about 0.2 and 0.3 and then steadily decreased. The maximum/minimum phenomenon was ascribed to the transition from an anisotropic rivulet flow to the film flow as the local flow density in this region increased faster than the available liquid trajectories.

An isotropic film flow was observed from calculations for the permeability equal to 0.8. In all cases it was found that the small scale flow distribution at the exit level was less uniform than that at the inlet level. This behaviour can never be predicted by the diffusion models to be dealt with subsequently.

The number of bonds connected to a site is closely related to the coordination number of the packing. Larson et. al [213] report, using data of other authors, that the product of the threshold permeability of the bond and the coordination number for three-dimensional lattices is 1.5. The mean number of the active bonds attached to a site in a cubic lattice at the threshold is thus one and half.

The utility of the percolation concept models rests also in the ability to evaluate quantitatively the small scale flow nonuniformities. A flow maldistribution factor as defined, for instance, by Bemer and Zuiderweg [188], whose value is zero for perfectly uniform flow and unity for complete mismatch, decreases in all cases with increasing irrigation rate. This is readily explained by the transition from the isolated rivulet flow to isotropic (film) flow.

Crine et al. [192] investigated also the case of liquid distribution in a bed irrigated by a line source on the top using the percolation model concept. In contrast to the above outlined procedure one cannot distribute the "active" property to individual bonds randomly. Crine [192] assessed the probability by the binomial density distribution function. The distribution of the active bonds under these circumstances will be correct provided that the radial liquid flow distribution is governed by the Gaussian binomial distribution function (which is a diffusional process) and provided that the radial spread is not affected by local irrigation rate (linear process).

The profiles obtained of the local irrigation rate indicated strong local irrigation rate fluctuations apart from those resulting from the large scale flow maldistribution due to the uneven inlet flow distribution. The maldistribution factor evaluated for this case decreased also with increasing mean irrigation density but the values for the line source feed were found to be higher than for the uniform initial feed.

Models representing the flow in the bed as a bundle of liquid streams which are split by particles of the packing have been proposed also by Hoek et al [206] and Albright [221]. The stream is assumed to be split by the particle into two equal halves and diverted horizontally. The direction of the horizontal diversion, either in parallel or perpendicularly to the previous split, is determined randomly. The streams reaching the wall are assumed to be reflected back. The number of channels thus created in the bed was compared with experimental data [206] and a good agreement was found.

In spite of the availability of powerful computers today the models based on percolation theory are still fairly time consuming. Recent attempts have been directed at reducing the required computational efforts by obtaining analytical expressions for the distribution utilizing, e.g. statistical models combined with the percolation process (Marchot et. al [211], Fox and Fan [204]), or utilizing the concept of maximum configurational entropy [198].

In general one can say that models based on the percolation theory represent potentially a very powerful tool to analyze the behaviour of trickle beds. They open the possibility of analyzing the column hydrodynamics, local flow rates, small scale maldistribution and permit direct modelling of liquid hold-up and mean thickness of the liquid film in the bed. Regrettably, they so far have not gained wide acceptance.

The reasons for this are probably the following: The size of lattices representing the bed must be fairly large thus contributing to the large computer times required. The details of the percolation principle employed in the algorithm are sometimes obscure and strongly affect the outcome of the modelling. The results are difficult to combine with mass and heat balances in order to analyze the effect of flow distribution on the transport phenomena and chemical reactions taking place within the bed. The large scale nonuniformities (effects of the column walls) and structural inhomogeneities of the bed have not been investigated. Nonlinear phenomena such as flow interactions have not been investigated.

4.2. THE DIFFUSION MODEL CONCEPT OF LIQUID FLOW IN A PACKED BED

As already mentioned the diffusion models fall into the category of pseudohomogeneous models. Accordingly, the concept on which the models are based is not liquid rivulets or flow paths but, instead, the density of irrigation. This quantity is defined as the volume of liquid crossing a unit area of column cross section per unit time. In fact the density of irrigation is thus the superficial liquid velocity.

From the models one obtains "local" values of the density of irrigation as a function of spatial coordinates. However, the term "local" refers again to an element of column cross section (sampling area) sufficiently large with respect to a single particle of the bed, yet sufficiently small with respect to the cross sectional area of the column.

The most highly probable and therefore diffusional character of the flow of liquid trickling down a packed bed was recognized by Tour and Lerman [167--169]. These authors utilized the following equation to describe the distribution of liquid in a radially unconfined bed irrigated by a point source

$$f'(r', z') = \frac{Q}{4\pi D z'} \cdot \exp\left[-\frac{r'^2}{4Dz'}\right] \tag{4-1}$$

where f' is the dimensional density of irrigation expressed as a function of the dimensional cylindrical coordinates r', z'.

Tour and Lerman [169] then obtained distribution of liquid in unconfined beds irrigated by area sources by integration on the assumption that the area of the liquid distributor consists of a number of point sources.

Cihla and Schmidt [53,217], however, recognized, using stochastic theory of Chandrasekhara [172], that the distribution of the density of irrigation from a point source as described above in an infinite medium represents the distribution of the probability that a fluid element descending vertically by a distance (in the above notation) z' will shift horizontally (not necessarily in the radial direction) by a distance (in the above notation) r'.

As may be seen this probability for a given packing is a function only of the vertical distance between the two cross sections in question. The effect of the packing itself is expressed in the coefficient of radial spread, D, an analog of the diffusion coefficient that, however, has the physical dimension of length.

With this probability density function Cihla and Schmidt [53] derived and showed that the distribution of the density of irrigation in a confined bed of packing is governed by the following differential equation written here in cylindrical coordinates r', z', φ as:

$$\frac{\partial f}{\partial z'} = D\left[\frac{\partial^2 f}{\partial r'^2} + \frac{1}{r'}\frac{\partial f}{\partial r'} + \frac{1}{r'^2}\frac{\partial^2 f}{\partial \varphi^2}\right] \tag{4-2}$$

The effect of the confining walls then enters the solution through proper formulation of the boundary condition. Cihla and Schmidt further assumed that the wall is impervious to the liquid and therefore the gradient of the density of irrigation in the direction perpendicular to the wall must equal zero. They presented a number of solutions for various types of liquid distributors (initial condition).

Of these it is worth reproducing here the three-dimensional solution for the case of a point source asymmetrically located at the dimensionless radial distance from the center r_1, expressed as a fraction of column radius, and the angular coordinate φ_1

$$f = 1 + 2 \sum_{n=0}^{\infty} \sum_{j=0}^{\infty} \frac{q_{nj}^2 J_n(q_{js} r_1)}{(q_{nj}^2 - n^2) J_n^2(q_{nj})} \cdot$$

$$\cdot \cos[n(\varphi - \varphi_1)] \cdot J_n(q_{nj}) \cdot \exp\left[-\frac{q_{nj} D z'}{R^2}\right] \tag{4-3}$$

where q_{nj} *are roots of*

$$J_n(q_{nj}) = 0 \qquad j = 1, 2, 3..... \tag{4-4}$$

and f is the dimensionless density of irrigation, expressed as a fraction of the mean density of irrigation.

4.2.1. The coefficient of radial spread of liquid

Let us rewrite now the diffusion equation of Cihla and Schmidt (4-2) into a dimensionless form for the case of axially symmetric flows:

$$H\left[\frac{\partial^2 f}{\partial r^2} + \frac{1}{r}\frac{\partial f}{\partial r}\right] = \frac{\partial f}{\partial z} \tag{4-5}$$

The radial and axial coordinates were scaled by the column radius R and the bed depth, h, respectively and H is a dimensionless group defined as

$$H = \frac{Dh}{R^2} \tag{4-6}$$

H thus may be considered to be either the dimensionless characteristic of the spreading ability of the packing or the dimensionless bed depth.

Had we scaled the spatial coordinates by the size of the packing element the coefficient on the left hand side of the equation would have had the form D/d_p. The inverse of this dimensionless group, namely, d_p/D, is an equivalent of the Peclet number for the radial spread of liquid in the packing.

The best way of determining the coefficient of radial spread, D, or some of its dimensionless forms, is to perform experiments so as to eliminate the effect of the walls containing the bed on the distribution of liquid.

This can be accomplished by experiments in which the liquid is fed into the bed as a thin stream and the depth of the packed layer, h, is not excessively large, measured in terms of the dimensionless group, H, so that no appreciable amount of liquid reaches the wall.

Under these circumstances the bed can be considered as being radially unconfined and the solution of Eq.(4-5) is given by Eq.(4-1).

Equally, one can utilize solutions of Eq.(4-5) with the boundary conditions characterizing a "perfectly dry" wall (the wall acts as a "sink" for liquid):

$$f = 0 \quad \text{for} \quad r = 1 \quad \text{and all} \quad z \tag{4-7}$$

or a wall "perfectly reflecting" the liquid back to the packing:

$$\frac{\partial f}{\partial r} = 0 \quad \text{for} \quad r = 1 \quad \text{and all} \quad z \tag{4-8}$$

In any of these cases the boundary condition introduces no additional parameter into the solution and if the above mentioned conditions are met (narrow axial irrigation of the packing on the top and a shallow bed of packing), it is immaterial if the boundary conditions are physically sufficiently well justified.

Solutions for all these boundary conditions and a hypothetical point source, or a disc source of radius r_1, can be found in the literature, for instance, Staněk and Kolář [2]. However, it should be born in mind that for the case of an unconfined bed or a perfectly dry wall part of the liquid appears always beyond the radius R, or beyond $r = 1$, and thus the material balance of liquid is violated. This error, however, will not be significant as long as the bed is shallow and the corresponding values of H are low.

The experimental determination of the parameter D, or H then consists in matching the computed profiles of the density of irrigation as a function of radial coordinate with experimental points.

The matching is particularly easy when one uses the concept of an unconfined bed irrigated by a central point source. For this case the solution in Eq.(4-1) can be linearized by using the log-quadratic coordinates and the corresponding value of H is obtained from the slope of the curve.

There is, however, a very definite drawback to using narrow central jets as liquid distributors for evaluation of the coefficient of the radial spread. In this case all the liquid is being fed onto a small area of the top surface so that the local densities of irrigation reach extremely high values. The voids of the packing are completely filled by liquid and the distribution follows a different mechanism.

A practical consequence is that optimum values of the coefficient of radial spread, found for various mean densities of irrigation of the packing, f_0, exhibit an apparent dependence on the mean density of irrigation [3,5] starting from a certain limiting value. This value can be set [3] at 0.0055 ms^{-1}. Below this limit, the value of D obtained from the narrow jet experiments is essentially constant and above this limit increases with the mean density of irrigation f_0.

An additional consequence of the flooding below the mouth of a narrow jet is that the values of D obtained differ for different depths of the bed with the same value of f_0.

At high mean irrigation rates the packing immediately below the inlet is not only completely filled by the liquid, but, part of the liquid actually cannot penetrate into the void available below the mouth of the nozzle and spreads radially over the top surface [17,180] of the bed (see Fig.4-1). The narrow jet thus effectively transforms into a disc whose radius r_1 (see Fig.4-1) depends on the mean irrigation rate and the geometry of the packing.

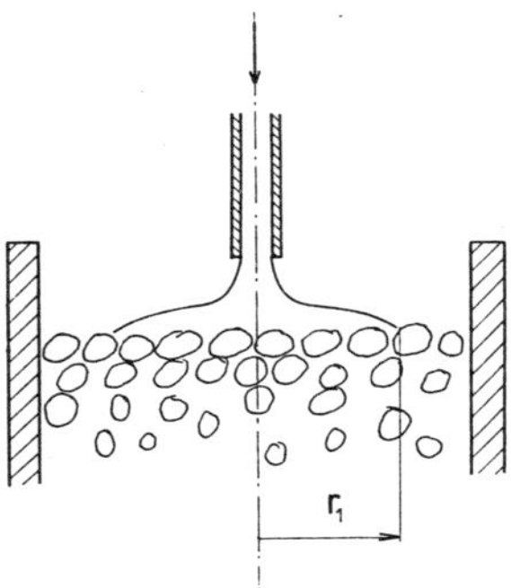

Fig.4-1
Sketch of liquid spreading on the top of the bed.

This phenomenon has been investigated experimentally by Staněk [17] and an attempt has been made to calculate the radius of the disc distributor transformed from the central jet by using the Ergun equation in which the pressure loss per unit length of the bed was replaced by the gravity force acting on the liquid. One can thus calculate the maximum superficial velocity of liquid in the pores of the packing driven by the gravity force and from this value in turn the radius to which the liquid jet must spread in order to reduce the initial density of irrigation to this limit.

The phenomenon was also investigated experimentally [17] by photographing the top surface of the bed irrigated by liquid containing a dark dye and these data were used to correct the coefficient in the dependence between the effective radius of the disc to which the jet transforms and the mean density of irrigation of the bed derived from the modified Ergun equation.

The results of experiments of the liquid distribution from a central jet, evaluated on the basis of a disc distributor [17] whose effective radius was computed as above, indicated that the optimum values of D no longer increase with the mean density of irrigation. In fact they were found slightly to decrease and, moreover, the dependence of the results of D on the depth of the bed practically vanished.

It was also encouraging to find that the values of D so obtained were essentially the same as those obtained from experiments with an actual disc distributor whose radius equalled one half of the column radius.

Nevertheless, it should be pointed out that the transformation of the jet into a disc distributor at the top of the packing is more difficult to predict for particles lacking spherical symmetry to a greater extent, such as e.g. Raschig rings.

Under such circumstances the top layer of particles, that is its structure, varies significantly from case to case and, in addition, the particles in this region usually lie with their axis horizontal. The spreading of liquid directly on the top of the bed (transformation of the jet to disc distributor) is then higly spatially oriented and the shape of the disc is strongly deformed.

Another method [17] of eliminating the effect of the spreading of liquid from a narrow jet directly on the top of the bed is to carry out measurements for two depths of the bed while using the distribution of liquid on the lower of the beds, fitted by any of the solutions mentioned above, as an initial distribution for the lower section of the bed whose depth equals the difference of the depths for the two measurements.

This method is essentially identical with that proposed by Bemer and Zuiderweg [219] and Zuiderweg et al. [206] who noted that the spreading of the jet can be computationally eliminated by increasing the true height of the bed by a small fictitious height, termed the "splashing height". The splashing height, however, behaved rather erratically [206] and was found to be higher for cylindrical Raschig rings and lower for more open types of the packing such as Pall rings and Intalox saddles.

The influence of local conditions of irrigation on the coefficient of radial spread has been investigated [19] on the basis of the equation that admits the dependence of D on the local conditions, namely

$$\frac{1}{r}\frac{\partial}{\partial r}\left[Hr\frac{\partial f}{\partial r}\right] = \frac{\partial f}{\partial z} \tag{4-9}$$

Upon integrating the individual terms of this equation in the limits between zero and the dimensionless radius r_1 with the weight $2r$ we obtain, after some manipulation, the following expression for the coefficient of radial spread of liquid, namely its value for the conditions of irrigation prevailing on the radius r_1:

$$D(r_1) = \frac{R^2}{h} \frac{\dfrac{\partial}{\partial z}\left[\int_0^{r_1} rf \, \mathrm{d}r\right]}{r_1 \left(\dfrac{\partial f}{\partial r}\right)_{r=r_1}} \tag{4-10}$$

The underlying experimental technique for evaluating $D(r_1)$ consists of measuring the radial density of irrigation profiles for gradually increasing depth of the bed. From these data one can easily evaluate the volumetric flow rates within the radii r_1 and the gradients of the density of irrigation. For better accuracy of the data processing the volumetric flow rates and the densities of irrigation are first smoothed by fitting suitable curves expressing their dependence on the radial and axial coordinate.

The results of processing a large number of data [19] according to the formula (4-10) for spheres and Raschig rings were plotted in the form of a chart of local values of the coefficient of radial spread in the density of irrigation versus gradient of the density of the irrigation domain.

The results for the packing of 20 mm spheres showed the existence of two regions: one for the low density of irrigation and low gradients and the other for high liquid densities and gradients. In these two regions D was constant but different one from another, with the lower value of D being in the upper region. The limit between these two regions was given by the value of the dimensional gradient about 10 (1/s).

In addition there existed another region of lower gradients and higher densities of irrigation where the values depended on local conditions. This last region was found particularly extensive for the packing of 15 mm and 25 mm Raschig rings. The contours of the constant D in this region were approximately diagonal in the density of irrigation versus gradient of the density of irrigation domain.

In fact all points available from the above experimental technique covered approximately a strip located also diagonally in this domain. From this it follows that for a given mean density of irrigation f_0 the local densities of irrigation and local gradients of the density of irrigation existing in reality in the bed are mutually correlated. Consequently, the same position of the contours of constant D in this domain means that the corresponding value of D for given irrigation of the bed is practically constant and one can thus safely use the linear model (4-5) instead of (4-9).

The results of the distribution of the density of irrigation so far discussed were obtained on the basis of experimental measurement of the density of irrigation profiles. These profiles can be obtained by collecting liquid at the bottom of the bed into individual chambers mostly in the form of honeycomb structures, as done e.g. by Jones [173], or a set of small diameter cylindrical cells, as done e.g. by Hoftyzer [174]. In most cases, however, the collecting device is formed by a set of concentric cylinders, see e.g. Baker et al [166], Farid and Gunn [193] and Staněk and Kolář [4,5].

In the latter case the total cross sectional area is divided into m basic annuli collecting the liquid and one can form [5] a total of $m_t = \sum_{i=1}^{m} i$ derived annuli by combining the neighbouring annuli. Among these combined annuli there will be, of course, some quite wide annuli. For the mean radius of these annuli, r_m, one can derive the following formula

$$r_m = \frac{2}{3}\left[r_e + \frac{r_i^2}{r_e + r_i}\right] \tag{4-11}$$

where r_e and r_i are external and internal radii of the derived annulus. The last expression was derived on the assumption that the distribution of the density of irrigation in region between r_i and r_e is linear. If the assumption of linearity is not well satisfied, as is the case with very wide annuli, such experimental points should be ommited from the set m_t.

A substantially different approach to evaluating the coefficient of radial spread, D, is experimenting with tracers. Consider a bed uniformly irrigated by liquid. Within a small central region (the central point source or a small radius disc distributor) the liquid carries a tracer. The concentration of the tracer is detected at the outlet of the bed as a function of the radial coordinate.

Provided that the controlling mechanism of radial mixing of the tracer is the radial spread of liquid itself one can write a balance on the tracer (dimensionless concentration c) in a form analogous to Eq.(4-5):

$$H\left[\frac{\partial^2(fc)}{\partial r^2} + \frac{1}{r}\frac{\partial(fc)}{\partial r}\right] = \frac{\partial(fc)}{\partial z} \tag{4-12}$$

where the coefficient of the radial spread, D, contained in the dimensionless group H, defined in Eq.(4-6), is the same as in Eq.(4-5).

This will be so provided that the contribution to the net radial spread of the tracer of molecular diffusion and turbulent mixing within liquid streams is negligible. The justification of these assumptions is readily acceptable and has been also confirmed experimentally [206].

The last equation can be arranged to yield [15,218]:

$$Hf\left[\frac{\partial^2 c}{\partial r^2} + \frac{1}{r}\frac{\partial c}{\partial r}\right] = f\frac{\partial c}{\partial r} - 2H\frac{\partial c}{\partial r}\frac{\partial f}{\partial r} \tag{4-13}$$

From this form it can be seen that if the initial distribution of liquid is uniform and the boundary conditions are such that the uniformity of the flow distribution is retained throughout the bed (no wall flow formation) the gradients of the density of irrigation are zero and the second term on the right hand side of the last equation vanishes. The equation may then factored by $1/f$ and changes to the same form as Eq.(4-5). The solutions for the distribution of the tracer in the bed are then formally the same as the solutions of the density of irrigation within the bed.

An experimental study to determine the coefficient of radial spread using this tracer technique [15] for spheres and Raschig rings of various size showed that optimum values of the coefficient of radial spread are essentially the same when one uses the linear solution (i.e. the solution of the concentration analog of Eq.(4-5)) or a numerical solution of Eq.(4-13).

However, the standard deviation of experimental data from the numerical solutions of the last equation was substantially lower, in some cases by more than 50 %. In case of this study [15] the liquid carrying the tracer was within a central disc of dimensionless radius $r_1 = 0.5$. Thus it can be concluded that the existence of non-zero radial density of irrigation gradients contributed non-negligibly to the radial spread of the tracer.

On the other hand, the results obtained of the coefficient of the radial spread were practically independent of the mean density of irrigation of the bed and the same regardless of the depth of the bed.

The tracer technique of determining the coefficient of radial spread offers an interesting possibility of estimating the number of rivulets trickling down the bed of packing [15,180]. If it is assumed that the number of rivulets does not vary along the depth of the bed, the χ^2 quantity testing the goodness of fit of the model distribution is proportional to this number. Since in contrast to the majority of the χ^2 test applications this number is not known, the goodness of fit test may be used in reverse to evaluate the number of rivulets from the goodness of fit data obtained by comparing the theoretical and experimental points and evaluating the χ^2 quantity.

The results of the estimate of the number of rivulets for packings of spheres and Raschig rings of various size [15] showed that the number of rivulets , N_r, at low mean densities of irrigation decreased with the packing diameter and the decrease obeyed approximately the following formula

$$N_r = \frac{3}{2}(1 - \varepsilon)\left(\frac{d_c}{d_p}\right)^2 \tag{4-14}$$

determining the number of spherical particles seen (cut) in a horizontal plane across the bed and derived on the basis of results of Debbas and Rumpf [45]. Acceptably, the number of rivulets at low densities of irrigation is proportional to the number of elements of the packing seen in a horizontal cross section of the bed and as such is determined by the geometry of the column and the packing.

The number of rivulets estimated on the basis of reverse goodness of fit at larger mean irrigation densities ($f = 0.006 \text{ ms}^{-1}$) was found to be independent of the particle size and in general the number of rivulets increased with the mean density of irrigation. Ultimately, however, the estimated number of rivulets exceeded the number of pour points of the distributor and hence the number of the rivulets brought onto the top of the packing. Since additional splitting of the rivulets within the bed is unlikely, the apparent increase of the estimated number of rivulets was ascribed to the mutual mixing of the rivulets. This phenomenon, of course, destroys the ability of the goodness of fit test to predict the number of the rivulets trickling within the bed.

From the above data [15] one can estimate the flow rate of a single rivulet which for low mean irrigation rates ($f_0 = 0.002 \text{ ms}^{-1}$) and Raschig rings amounted to $0.27 \text{ cm}^3\text{s}^{-1}$. Porter et al [180] found the mean values between 0.2 and $0.6 \text{ cm}^3\text{s}^{-1}$ the higher values being for increased mean irrigation rates.

Stikkelman and Wessenlingh [207] developed a mathematical model that in fact opens up the possibility of calculating the coefficient of radial spread of liquid, D, from the known maldistribution factor. The model [207] simulates a random packing by a large number of small boxes. A constant fraction of liquid (liquid split factor) that enters the box is diverted and leaves the box through one of the randomly picked sides.

Similarly, a constant fraction of the gas flow entering the box countercurrently through the bottom and the sides leaves the box (gas split factor) evenly through all sides except the one through which the liquid enters the given box. The remaining part of the gas leaves through the top of the box.

Calculations on the basis of this model [207] revealed that the maldistribution factor, Mf, defined both for gas and liquid as

$$Mf = (f_{\text{local}} - 1)^2/n \tag{4-15}$$

where f_{local} is the local value of the dimensionless density of irrigation for liquid or dimensionless superficial velocity of gas, depends for gas on the gas split factor only and for liquid, similarly, on the liquid split factor only. The liquid split factor was further found to be directly proportional to the product Dh_b where h_b is the height of the box of the model packing (equal d_p). This in turn offers the possibility of calculating the coefficient of radial spread of liquid from the known maldistribution factor Mf.

The above definition of the maldistribution factor, *Mf*, is due to Groenhof [196]. In its original form it presumes summation over all sampling locations (i) and time (j). Thus $n = i.j.$

4.2.2. Numerical values of the coefficient of radial spread and correlations

Tour and Lerman [168] attempted to predict the relationship between the coefficient of radial spread and the characteristic dimension of the packing, d_p. They compared the results of the fundamental solution (normal distribution) in Eq.(4-1) with the binomial distribution and found that the product (Dh) should be proportional to the square of the horizontal displacement per single vertical step multiplied by the number of vertical steps.

If we assume that the horizontal displacement per single vertical step is proportional to the particle size d_p and that the number of vertical steps in proportional to h/d_p we obtain the following proportionality for the coefficient of radial spread

$$D = k_1 d_\mathrm{p} \tag{4-16}$$

For structured packings the dimension in the horizontal direction, d_r, which determines the size of the horizontal displacement, may be substantially different from the dimension in the vertical direction, d_z, that determines the number of vertical steps. A consideration analogous to the above [206] yields then the following proportionality for the coefficient of radial spread

$$D = k_2 \frac{d_r^2}{d_z} \tag{4-17}$$

In spite of the predicted proportionality between the coefficient of radial spread and the size of the packing, some of the experimental studies indicated that D should rather be a constant with specific values for each packing type. See e.g. Staněk and Kolář [3], for spheres and Raschig rings, or Kolomazník et al. [184] for catalyst pellets.

This finding however can be attributed to the fact that the above measurements [3] were carried out with particles as small as 5 mm. Under such circumstances the stagnant hold up of liquid in the bed is large, particularly when experiments are carried out with water and the large volume of liquid bridges between individual particles obscures the effect of the geometry of the packing. Other studies support the proportionality to the packing size, see e.g. Staněk and Kolář [15], Dutkai and Ruckenstein [181], Hoftyzer [174], Hoek et al [206], Brignole [189] and others.

The concept of the peculiar behaviour of small particles in relation to the static hold up seems to be supported also by the finding [15] that an almost stepwise increase of the value of D was observed for spheres and Raschig rings for values of d_p about 10 mm. This lead Onda et al [54] to the formulation of the following correlation relating the coefficient of radial spread to the square root of d_p for Raschig rings and Berl saddles:

$$D = 0.169\ d_p^{0.5} \qquad (D \text{ and } d_p \text{ in centimeters}) \tag{4-18}$$

However, Porter and Jones [215] observed that eliminating the results with 4 and 8 mm particles Onda et al [54] data yield the following result

$$D = 0.122\ d_p \tag{4-19}$$

Note that for particles larger than 10 mm the difference of results will not be large.

Hoftyzer recommends the value of the proportionality constant k_1 in Eq.(4-16) for rings and saddles as equal to 1/7, while Hoek et al [206] recommend $k_1 = 0.12$ for Raschig rings and Intalox saddles. This value also exhibits aproximately the correlation of Brignole et al [189] for ceramic Raschig rings. It is interesting to note that the inverse of the Hoek et al [206] value equals approximately $(1/k_1) = 8$ which is a value of the Peclet number representing the effective radial diffusivity of mass in the bed. Hence the radial spread of liquid is the dominating mechanism of radial mixing for the mass.

The same authors [206] evaluated the constant k_2 in Eq.(4-17) for the Mellapak (Sulzer) packing to be $k_2 = 0.22$.

For packings and catalyst pellets usually used the effect of the shape does not seem to be very marked as reported by many authors, e.g. Bemer et at [188], Kolomazník et al [184], Soukup et al [183]. This would indicate that the proportionality constant k_1 in Eq.(4-16) should not change dramatically, but Porter and Jones [173] report that D tends to decrease in the series Raschig rings, Berl saddles, Pall rings.

There may be a significant effect caused by the method of packing (dumping) in the case of: highly non-spherical particles when the principal dimensions are rather different; or where the efficiency of individual facets of the particle to transfer liquid horizontally is different.

Perktold [171] investigated the spreading of liquid in beds of 12 mm Raschig rings prepared in three different ways:

a) central - packing deposited at the axis of the bed,

b) neutral - packing spread by parts in horizontal layers,

c) peripheral - packing deposited toward the wall.

Staněk and Kolář [3] processed their data to obtain D and found that the coefficient of radial spread decreased in the sequence central (a), neutral (b), peripheral (c).

The least spreading ability for the peripheral packing means that the rings were oriented predominantly with their axes vertical; in this position their ability to divert liquid radially is the least. In contrast, the largest spreading ability for the central packing means that the rings rested with their axes predominantly horizontal, in which position the cylindrical surface is offered for radial transport of liquid.

It appears that the quality of the material of construction of the particles does not affect their liquid spreading properties. Porter et al [138] came to this conclusion for ceramic and steel particles. Kolomazník et al [184] report for catalyst particles that the porous pellets in comparison with non-porous particles have about 20 % lower spreading coefficient. Bemer and Zuiderweg [188,219] conclude that the wettability of the surface of particles has no effect on the coefficient of radial spread. This is, however, not true for new packings. Nakamura et al [205] report that a new packing should be wetted for several hours before use to avoid severe maldistribution.

The spreading of liquid is apparently affected also by the physical properties of liquid. Staněk and Kolář [5] found the coefficient of radial spread to be lower for liquids such as aqueous solution of detergent, glycerol and isobutanol compared with the values of D for water and the same packing. This was attributed to the decreased surface tension of these liquids. Also Hoftyzer [174] reported that a decrease of the surface tension from the value for water to about 42 dyne cm^{-1} had no effect but a further decrease to 32 dyne cm^{-1} considerably improved the distribution.

Onda et al [54], in fact, modified their correlation (4-18) to account for the effect of surface tension of the irrigating liquid to the form

$$D = 0.00231 d_p^{0.5} \sigma \qquad (D,\ d_p \text{ in [cm]},\ \sigma \text{ in [dyne cm}^{-1}\text{]}) \tag{4-20}$$

This correlation supports the previous conclusions that decreasing surface tension decreases D. The same authors report that the effect of liquid viscosity is negligible.

The effect of gas velocity on the spreading of liquid is regarded as being mostly negligible in the no-interaction region. Dutkai and Ruckenstein [55] detected no effect of gas velocity up to 70% of the flooding velocity. Hoftyzer [174] found no effect up to the loading point.

More recent studies, however, indicate that the effect of gas velocity should be considered. Kouri and Sohlo [208] report that above the gas mass velocity of 2 $kg\,m^{-2}s^{-1}$ the maldistribution in the packing of Pall rings and BX (Sulzer) packing increased. In particular, Nakamura et al.[205] found by simultaneous measurement of gas and liquid distribution that at higher flow rates gas preferred the wall region pushing the liquid toward the axis. This seems to suggest that the primary effect of the increased void fraction near the wall is that on the gas flow rather than the liquid flow.

El-Din et al.[187] found that the increased gas velocity tends to flatten the liquid flow profiles. This points again to the primary role of gas flow in affecting the flow distribution, but, according to Stikkelman and Wesselingh [207], gas flow maldistribution is small compared to liquid flow maldistribution while there are significant differences for individual packings.

The role of gas on the flow distribution under the two phase flow can therefore be probably summarized as follows:

At low irrigation rates the dominating effect is that of liquid. This means that eventual nonuniformities of liquid flow would bring about (mostly through variable local liquid holdup) the nonuniformities of the gas flow.

At higher liquid irrigation rates under both the co- and counter-current flow regime, when detectable gas/liquid interaction appears (the loading point for the counter-current flows), the gas flow seems to suppress the small scale maldistribution, but on the large scale the controlling effect is that of the gas. In other words, there would be smaller differences in the irrigation rates between individual particles, especially those that are not far apart, but, on the large scale, the gas would displace the liquid from the regions of lower resistance (the wall region) so as to flow preferentially in this area.

In the absence of any information about a particular type of the particles of packing or catalyst pellets it is recommended to use Eq.(4-16) with the coefficient k_1 equal to 1/8.

4.2.3. The wall flow

In the previous paragraphs we have examined the characteristics of the spreading of liquid in the bed which are best investigated if the bed is irrigated by a central jet or discs of small radius compared to the diameter of the columns. This way one can avoid the effects of the containing walls of the column.

In practical situations, however, the liquid is brought to the top of the bed by distributors of differing construction and the deviations of the flow of liquid from uniformity then appear for at least two reasons. One is that the distributor does not irrigate the top of the bed uniformly, and, second, that the liquid reaches the containing wall of the column where the distribution mechanism is generally different from that within the bed. There is ample evidence in the literature that there is a tendency of liquid to accumulate on the wall and to form here what is termed the wall flow [194].

Nevertheless, the problem of liquid distribution in a bed, including the effect of the containing walls, can be attacked as a solution of Eq.(4-2), for three dimensional flows, or Eq.(4-5) in the dimensionless form for two dimensional axisymmetric problems. The problem then reduces to one of formulating a suitable boundary condition characterizing the exchange of liquid between the bulk of the packed bed and the wall region.

In the previous paragraphs we have already noted two types of boundary conditions viewing the wall as a perfect sink for liquid, Eq.(4-7) or as a wall perfectly reflecting the liquid, Eq.(4-8). For practical situations where a non--negligible amount of liquid reaches the wall these boundary conditions are unrealistic and the corresponding solutions provide počor match with experimental data.

An attempt to formulate a boundary condition that would express the fact that the real behaviour of the wall, as far as the liquid distribution is concerned, is somewhere between the perfect sink and the perfect reflector, was made by Porter and Jones [173]. In order to express the ability of the wall to return the liquid back to the packing they assumed that the density of irrigation in the immediate vicinity of the wall is proportional to the wall flow. Using, in contrast to the original paper [173], a dimensionless notation this boundary condition reads

$$f(r = 1, z) = K\, W(z) \tag{4-21}$$

$f(r = 1,z)$ designates the dimensionless density of irrigation of the bed near the wall. $W(z)$ designates the dimensionless wall flow, that is the volumetric flow rate of liquid down the wall as a fraction of the total volumetric flow rate in the bed $\pi R^2 f_0$. K is the dimensionless wall effect factor.

This boundary equation in effect states, that at the bottom of an infinitely deep bed the wall flow and the density of irrigation (for any value of r) are at equilibrium. From a simple balance on the liquid one can therefore determine that for an infinitely deep, or in practice sufficiently deep (judged from the value of the criterion H) bed, the density of irrigation is

$$f(r, z = \infty) = \frac{K}{1 + K} \tag{4-22}$$

and the wall flow is given by

$$W(\infty) = \frac{1}{1 + K} \tag{4-23}$$

These equations present convenient means of determining the values of the wall effect factor from the values of the wall flow and the density of irrigation obtained on sufficiently deep beds.

However, Staněk and Kolář [4] evaluated the wall effect factor by fitting experimental density profiles to the solutions of Eq.(4-5) using the boundary condition Eq.(4-21) for the case of the uniform initial wetting of the bed and for the case when the liquid on the top of the bed is brought onto the wall. These fittings were carried out for experiments on several depths of the packing and it was found that the wall effect factor decreases with increasing depth of the bed for uniform initial irrigation and, on the other hand, increases with the depth of the bed if the liquid is brought initially onto the wall.

For high values of the bed depth both trends seem to converge to the same value which supports the idea of "equilibrium" between the wall flow and the density of irrigation but only for infinitely deep beds. Since the uniform distributor generally induces flows toward the wall and the wall distributor generally induces flows toward the axis of the bed it can be concluded that in the course of the distribution process the liquid shows preference for that side from which the flow begins or originates.

The rate of liquid transfer between the wall and the bed is therefore always less than that determined by the equilibrium value of K which is in direct violation of the assumption that the flow near the wall is proportional to the wall flow.

Templeman and Porter [175,57], Prchlík et al. [185] and Dutkai and Ruckenstein [55] also expressed the opinion that the boundary condition (4-21) is an oversimplification.

To circumvent these deficiencies Kolář and Staněk [1] proposed a boundary condition in the following dimensional form

$$-D\left(\frac{\partial f'(r', z')}{\partial r'}\right)_{r'=R} = \beta\left[f'(r' = R, z') - \gamma\, W'(z')\right] \tag{4-24}$$

This condition expresses the net radial flux of liquid to the wall (the left hand side term) as a product of the coefficient of transfer of liquid to the wall, β, and the driving force for this process. For the driving force the authors took the difference between the density of irrigation near the wall and the wall flow. However, another coefficient, γ, was introduced here to account for the existence of the "equilibrium" conditions between the wall flow and the density of irrigation at large depths of the bed, similarly as done by Porter and Jones [173]. This form of the boundary condition in fact suggests that "diffusion" of liquid toward the wall is not the rate determining step in establishing the equilibrium between the wall flow and the density of irrigation near the wall on a given vertical level.

In the more convenient dimensionless form the last boundary conditions reads [2]

$$\left(\frac{\partial f}{\partial r}\right) = B(f - CW);\ r = 1 \tag{4-25}$$

Later Dutkai and Ruckenstein [55] also formulated a boundary condition with somewhat different arrangement of the coefficients but the same mathematical meaning.

Onda et al. [54] proposed the following boundary condition in the dimensional form written as:

$$-2\pi RD\left(\frac{\partial f'}{\partial r'}\right)_{r'=R} = k(W'_{eq}(z') - W'(z')) \tag{4-26}$$

where $W'_{eq}(z')$ designates the wall flow at equilibrium with the density of irrigation in the vicinity of the wall and $W'(z')$ is the current local value of the wall flow. The authors solved then Eq.(4-5) numerically with the above boundary condition and they results differed very little from those obtained on the basis of the boundary condition (4-21) or (4-25).

The boundary condition (4-25) contains two dimensionless parameters B and C and the wall flow is regarded as a separate entity different from the flow over the surface of the packing. The magnitude of the wall flow, of course, varies with the axial coordinate. The balance on the liquid, both within the bed and on the wall, gives the following relationship in the dimensionless form

$$\frac{\mathrm{d}W}{\mathrm{d}z} = -2H\frac{\partial f}{\partial r} \quad \text{for} \quad r = 1 \tag{4-27}$$

4.2.4. The distribution of liquid under various initial distributions and under the effect of the wall flow

The boundary condition (4-25) was found by its authors to fit best the experimental data [4]. Other authors, e.g. Prchlík et al [185] and Furzer [203], found the same.

Table 4-1 summarizes solutions of Eq.(4-5) with the boundary condition given by Eq.(4-25) for various types of liquid distributors determining the initial condition. The other boundary condition at the axis of the bed was obviously the condition of axial symmetry, i.e. a zero gradient of the density of irrigation in the radial direction.

q_n appearing in the tabulated solutions are roots of the following equation

$$\left(\frac{2C}{q_n} - \frac{q_n}{B}\right) J_1(q_n) + J_0(q_n) = 0 \tag{4-28}$$

ordered in the sequence according to the increasing magnitude.

In addition one can also write [12] a solution for an arbitrary, axially symmetric initial distribution, given by a function g(r), in the following form

$$f = \sum_n A_n J_0\left(\frac{r}{R} q_n\right) \exp\left[-q_n^2 \frac{Dz'}{R^2}\right] \tag{4-29}$$

where q_n are again the roots of Eq.(4-28) and A_n are obtained by the following integration using the known function $g(r)$:

$$A_n = \frac{2\left(\frac{q_n^2}{B} - 2C\right)^2}{\left[\left(\frac{q_n^2}{B} - 2C\right)^2 + q_n^2 + 4C\right] J_0^2(q)} \int_0^1 rg(r)\, J_0(q_n r)\, \mathrm{d}r \tag{4-30}$$

The easiest way of performing the integration is to expand the function $g(r)$ into a series of Bessel functions and utilize the orthogonality of the Bessel functions.

The zero-th term of the series in Eq.(4-29) is always equal

$$A_0 = \frac{C}{1 + C} = f(r, z = \infty) \tag{4-31}$$

Table 4-1

Solutions of Eq.(4-5) with Eq.(4-25) under selected initial conditions

Central point source $f = \begin{matrix} \nearrow \infty & \text{for} \quad r = 0 \\ \searrow 0 & \text{for} \quad 0 < r \leqq 1 \end{matrix}; \quad 2\int_0^1 rf\,\mathrm{d}r = 1$

$$f = \frac{C}{1 + C} + \sum_n \frac{[(q_n^2/B) - 2C]^2\, J_0(q_n r) \exp\,[q_n^2 zH]}{\{[(q_n^2/B) - 2C]^2 + q_n^2 + 4C\}\, J_0^2(q_n)}$$

$$W = \frac{1}{1 + C} - \sum_n \frac{2[(q_n^2/B) - 2C] \exp\,[-q_n^2 zH]}{\{[(q_n^2/B) - 2C]^2 + q_n^2 + 4C\}\, J_0(q_n)}$$

Disc distributor $$f = \left(\frac{1}{r_1}\right)^2 \quad \text{for} \quad 0 \leqq r \leqq r_1; \quad f = 0 \quad \text{for} \quad r_1 < r \leqq 1$$

$$f = \frac{C}{1 + C} + \frac{1}{r_1} \sum_n \frac{2[(q_n^2/B) - 2C]^2\, J_0(q_n r)\, J_1(q_n r_1) \exp\,[-q_n^2 zH]}{\{[(q_n^2/B) - 2C]^2 + q_n^2 + 4C\}\, q_n\, J_0^2(q_n)}$$

$$W = \frac{1}{1 + C} - \frac{1}{r_1} \sum_n \frac{4[(q_n^2/B) - 2C]\, J_1(q_n r_1) \exp\,[-q_n^2 zH]}{\{[(q_n^2/B) - 2C]^2 + q_n^2 + 4C\}\, q_n\, J_0(q_n)}$$

Annular distributor $$f = \frac{1}{r_2^2 - r_1^2} \quad \text{for} \quad r_1 \leqq r \leqq r_2; \quad f = 0 \begin{matrix} \nearrow 0 \leqq r < r_1 \\ \searrow r_2 < r \leqq 1 \end{matrix}$$

$$f = \frac{C}{1 + C} + \frac{1}{r_2^2 - r_1^2} \sum_n \frac{2[(q_n^2/B) - 2C]^2\, J_0(q_n r)\, [r_2\, J_1(q_n r_2) - r_1\, J_1(q_n r_1) \exp\,[-q_n^2 zH]}{\{[(q_n^2/B) - 2C]^2 + q_n^2 + 4C\}\, q_n\, J_0^2(q_n)}$$

$$W = \frac{1}{1 + C} - \frac{1}{r_2^2 - r_1^2} \sum_n \frac{4[(q_n^2/B) - 2C]^2\, [r_{2\,1}(q_n r) - r_1\, J_1(q_n r_1) \exp\,[-q_n^2 zH]}{\{[(q_n^2/B) - 2C]^2 + q_n^2 + 4C\}\, q_n\, J_0(q_n)}$$

Circular distributor $$f = \begin{matrix} \nearrow \infty & \text{for} \quad r = r_1 \\ \rightarrow 0 & \text{for} \quad 0 \leqq r < r_1 \\ \searrow 0 & \text{for} \quad r_1 < r \leqq 1 \end{matrix}\; ; \qquad 2 \int_0^1 r f \,\mathrm{d}r = 1$$

$$f = \frac{C}{1 + C} + \sum_n \frac{[(q_n^2/B) - 2C]^2\, J_0(q_n r)\, J_0(q_n r_1) \exp\,[-q_n^2 zH]}{\{[(q_n^2/B) - 2C]^2 + q_n^2 + 4C\}\, J_0^2(q_n)}$$

$$W = \frac{1}{1 + C} - \sum_n \frac{2[(q_n^2/B) - 2C]\, J_0(q_n r_1) \exp\,[-q_n^2 zH]}{\{[(q_n^2/B) - 2C]^2 + q_n^2 + 4C\}\, J_0(q_n)}$$

Wall distributor $$f = 0 \quad \text{for} \quad 0 \leqq r \leqq 1; \quad W = 1$$

$$f = \frac{C}{1 + C} - C \sum_n \frac{2[(q_n^2/B) - 2C]\, J_0(q_n r) \exp\,[-q_n^2 zH]}{\{[(q_n^2/B) - 2C] + q_n^2 + 4C\}\, J_0(q_n)}$$

$$f = \frac{1}{1 + C} + \sum_n \frac{4C \exp\,[-q_n^2 zH]}{[(q_n^2/B) - 2C]^2 + q_n^2 + 4C}$$

Uniform distributor $\qquad f = 1 \quad \text{for} \quad 0 \leqq r \leqq 1$

$$f = \frac{C}{1 + C} + \sum_n \frac{2[(q_n^2/B) - 2C]\, J_0(q_n r) \exp[-q_n^2 zH]}{\{[(q_n^2/B) - 2C]^2 + q_n^2 + 4C\}\, J_0(q_n)}$$

$$W = \frac{1}{1 + C} - \sum_n \frac{4 \exp[-q_n^2 zH]}{\{[(q_n^2/B) - 2C]^2 + q_n^2 + 4C\}}$$

and, as also shown in the last equation, determines the dimensionless density of irrigation on an infinite depth of the packed bed. The fact that the same equilibrium relationship between the density of irrigation and the wall flow is ultimately reached independently of how the liquid is fed initially on the top has been long well established by e.g. Baker et al. [166], Porter and Jones [173], Porter [179], Onda et al [54].

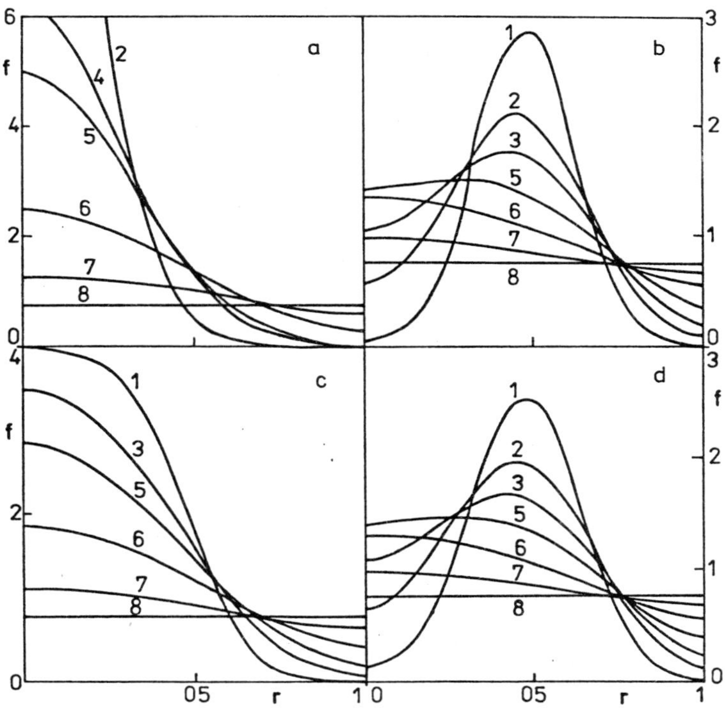

Fig.4-2
Radial profiles of the dimensionless density of irrigation computed for B=6.82, C=3.12 and a) Central point source, b) Circular distributor of $r_1 = 1/2$, c) Disc distributor of $r_1 = 1/2$, d) Annular distributor of $r_1 = 3/8$ and $r_2 = 5/8$. Curves 1 - 8 computed for H=0.01; 0.02; 0.03; 0.04; 0.05; 0.1; 0.2 and ∞ respectively.

Some of the solutions obtained are shown in Fig.4-2. From these figures, as well as from the inspection of the above solutions, it is seen that even for the case of initial uniform wetting of the bed deviations from the uniformity of the flow will appear. In addition, these solutions indicate that after a sufficiently long depth of the bed the effect of the initial distribution becomes insignificant and the dominating large scale maldistribution effect is that of the wall flow.

In a somewhat different form the distribution of the density of irrigation is shown in Figs (4-3) and (4-4). The left hand vertical edge of the frame always represents the axis of the bed while the right hand side edge the wall of the column. In the first of these two figures the bed is irrigated uniformly on the top, as it is indicated by the arrows. Individual contours are lines of constant density of irrigation and it is seen that the regions near the wall are initially depleted faster of the liquid in order to form the wall flow. The tilting of the contours of the constant f also confirms the gradual build up of the wall flow.

In view of the specific value of the parameter C for which the calculation was

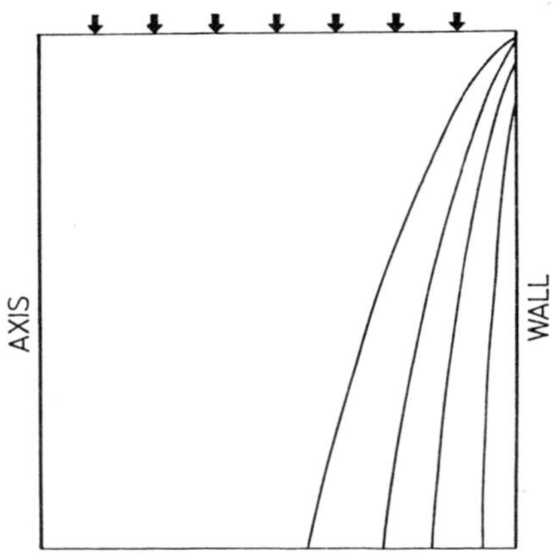

Fig.4-3
Contours of constant density of irrigation in the r-z plane computed for a uniformly irrigated bed and $C=2$, $B = 6.7$ and $H = 0.05$. Curves from left to right are contours of dimensionless density of irrigation equal 0.9; 0.8; 0.7; 0.6.

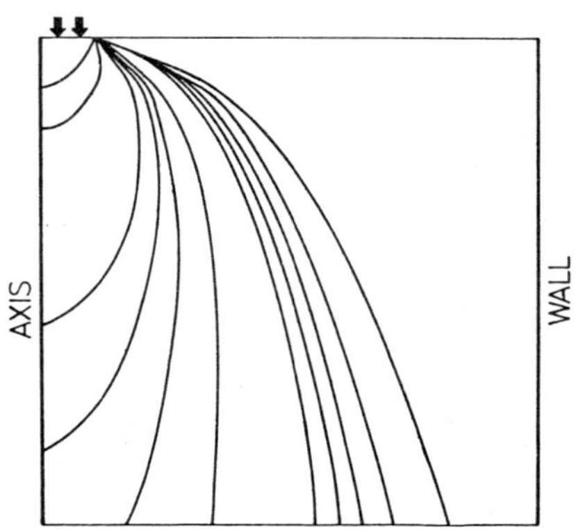

Fig.4-4
Contours of constant density of irrigation in the r-z plane computed for a narrow jet ($r_1 = 0.1$) distributor and $C = 2$, $B = 6.7$ and $H = 0.05$. Curves from left to right are contours of dimensionless density of irrigation equal 50; 30; 10; 7; 5; 3; 1; 0.7; 0.5; 0.3; 0.1.

made, $C = 2$, the ultimate equilibrium value of the dimensionless density of irrigation is 0.66. Accordingly, the contours of constant f for values higher than 0.66 would finally disappear into the axis while the contours for values lower than 0.66 would disappear into the wall. The type of maldistribution shown in this figure is typical for the columns under the effect of the wall flow.

The case depicted in Fig. (4-4) represents the bed irrigated by a narrow central jet (see the arrows on the top edge of the frame). The figure again shows a relatively short section of the bed below the feed (given by $H = 0.045\,6$). The maldistribution in this case is largely due to the nonuniformities of the feed. Accordingly, the region around the axis of the bed is excessively irrigated while the wall region is almost dry. For the same reason and in spite of the tendency of liquid to spread toward the wall there is no appreciable amount of liquid flowing down the wall. This would, of course, occur for greater depths of the bed (greater H).

There is no doubt that even if there are no chemical or thermal processes taking place in the bed, the flow inhomogeneities as shown must at least influence the hydrodynamics of the column.

4.2.5. The determination and numerical values of the boundary condition parameters C and B

A logical and the simplest method of determining the parameter C is to utilize the fact that this parameter controls the equilibrium wall flow and equilibrium density of irrigation at infinite depth of the bed. This means, for instance, Eq.(4-31) should be utilized. This approach has been used by a number of authors, as e.g. Staněk and Kolář [2,4,13], Dutkai and Ruckenstein [55], Porter and Jones [173] and Porter [57].

In some cases, however, there is no possibility of experimenting on a bed of such depth that it could be regarded as infinite. For this case there is an alternative approach [13] if we note that the distributions of the dimensionless density of irrigation (see Table 4-1) in a bed irrigated on the top uniformly, f_U, and the bed where all liquid on the top is fed onto the wall, f_W, can be written in the following form

$$f_U = \frac{C}{1 + C} + g(r,z) \tag{4-32}$$

$$f_W = \frac{C}{1 + C} - Cg(r,z) \tag{4-33}$$

where the function $g(r,z)$ represents the series of the Bessel functions. Simple algebra yields from the last two equations the following expression for the parameter C of the boundary condition

$$C = \frac{f_W}{1 - f_U} \tag{4-34}$$

For the success of this method of evaluation it is essential that both experiments, i.e. with the uniform and the wall flow distributor, be carried out on the same bed without repacking. The changes due to the method of packing could cause significant errors. The analysis further showed [13] that the accuracy of the last expression is better for greater depth of the packing as the contribution of the function $g(r,z)$ decreases.

Even better improvement is achieved when, instead of substituting individual values of the density of irrigation into the formula (4-34), we derive an analogous expression for the dimensionless volume flow rates $M(r_1, r_2)$ between the dimensionless radii r_1, r_2 for the two types of distributors, distinguished by superscripts [13]:

$$C = \frac{M^{w}(r_2, r_1)}{r_2^2 - r_1^2 - M^{u}(r_1, r_2)} \tag{4-35}$$

where

$$M^{w} = \int_{r1}^{r2} 2r\, f_W \, dr \tag{4-36}$$

$$M^{u} = \int_{r1}^{r2} 2r\, f_U \, dr \tag{4-37}$$

This approach in effect represents a certain smoothing procedure of individual experimental densities of irrigation. Also, it can be used together with an optimizing procedure to process data [13] on the dimensionless flow rates between the wall and the set of concentric cylinders. In this case $r_2 = 1$ and r_1 is the variable dimensionless radius of individual concentric collecting cylinders. Plotting then $(1 - r_1^2 - M^{u}(r_1, 1))$ versus $M^{w}(r_1, 1)$ for individual r_1 should give a straight line through the origin with the slope equal $1/C$.

From this formula it is seen again that if there is no tendency in the column to build up wall flow, the dimensionless flow rate between radii r_1, r_2 equals $r_2^2 - r_1^2$ and hence the corresponding value of the criterion C of the boundary condition is infinity.

Once the value of the dimensionless criterion C of the boundary condition is known, the other dimensionless criterion B can be obtained e.g. by fitting experimental density of irrigation profiles with the corresponding solution for the given initial type of distribution.

An alternative way is to substitute Eq.(4-27) into the boundary condition (4-25) and assume that for shallow beds irrigated by the wall distributor the density of irrigation near the wall, $f(r = 1, z)$, is negligibly small. The equation can be then integrated with the following result

$$W_W = \exp\left[-2BCH\right] \tag{4-38}$$

Similarly, for the uniform initial irrigation one can assume that for shallow beds $f = 1$ and the magnitude of the wall flow is negligible and obtain from integration of the same equation that

$$W_U = 2BH \tag{4-39}$$

The build up of the wall flow for the uniform initial distribution, as well as its reduction at the wall initial distribution, however, takes place fairly rapidly and therefore only very shallow beds, not exceeding a few particle diameters, may be used in order to meet the simplifying assumption leading to the last two expressions.

In general, experimentation on very shallow beds is not advisable for they are in the region of structural inhomogeneities (see chapter 2). In addition, relatively few contacts exist between the wall and the packing and hence the data will display considerable scatter depending on the method of packing.

These difficulties can be partly circumvented by assuming [13] that on shallow beds the density of wetting near the wall, $f(r = 1,z)$ may be put approximately equal to $(1\text{-}W)$. Integrating again the boundary condition (4-25) after substituting from Eq.(4-27) with the above simplification we obtain that

$$W_W = \frac{1}{1 + C} + \frac{C}{1 + C}\exp\left[-2B(1 + C)\,H\right] \tag{4-40}$$

$$W_U = \frac{1}{1 + C} - \frac{1}{1 + C}\exp\left[-2B(1 + C)\,H\right] \tag{4-41}$$

It is interesting to note that the last two approximate solutions provide physically correct values of the wall flow for infinitely deep beds, which is not the case for the simplified solutions (4-38) and (4-39). Even in this case, however, the experiments have to be carried out on beds not excessively deep in order to retain a reasonable agreement between the complete analytical solution and the approximate solution for the wall flow.

In order to avoid cumbersome expressions for the wall flow given by the analytical solution of the model (4-5) some authors, as e.g. Prchlík et al.[186] and Brignole et al.[189], proposed empirical correlations to express the dependence of the wall flow on the depth of the bed in the form similar to the previous equation

$$\frac{W}{W(\infty)} = 1 - \exp\left[-bH\right] \tag{4-42}$$

The parameter b ranges, according to Prchlík, between 0.39 and 0.66 depending on the type of particles.

As already noted the evaluation of B requires prior knowledge of the parameter C. A parametric sensitivity analysis has shown [13] that the sensitivity of B to an error in C is the least for low layers of the packing. This can be readily explained by the fact that the coefficient B characterizes the mechanism of the wall flow build up or decay and, understandably, experiments on deep beds carry little information about the dynamics. On the contrary, deep beds largely carry the information on the ultimate equilibrium value of the wall flow, a factor that is expressed by the parameter C. Large values of C, in addition, mean only a weak effect of the wall flow. Consequently, if any of the two parameters of the boundary conditions B, or C becomes large the other loses its importance.

An interesting approach was used by Zarzycki [56] and Hobler [181] to determine the volume of liquid flowing down the surface of the packing and the surface of the wall, and, hence, in effect the parameter C.

These authors modelled the packing by a zig-zag vertical surface P_{p} wide with the slope of individual facets being α. The total surface area of the model surface equalled the surface area of the packing. Similarly, the wall was represented by a flat vertical surface P_{w} wide with its total surface being equal to the surface area of the wall. Assuming laminar flow of liquid on both surfaces, an expression was derived for the total energy of liquid per unit height as being equal to the sum of the kinetic energy and surface energy of liquid on the packing and the wall.

Several situations can now be envisaged leading to the minimum energy dissipation:

a) When the Reynolds number (Re) for the liquid on the packing and the wall are both below the critical value so that neither surface is completely covered by liquid. In this case an inequality results containing the contact angles of wetting of the material of the wall and the packing. Depending then on which of the inequality signs is valid the liquid will exclusively flow either on the packing or on the wall.

b) When the situation is such that only the wall is thoroughly covered by the liquid film (Re for the wall above the critical value). Then there is a constant magnitude of the dimensional wall flow and all additional liquid after increasing the total liquid rate (mean density of irrigation) trickles down the surface of the packing.

c) When the situation is such that only the packing is thoroughly covered by the liquid film (*Re* for the packing above the critical value). Then there is a constant dimensional density of irrigation of the packing and all additional liquid, after increasing the total liquid rate (mean density of irrigation) increases the magnitude of the wall flow.

d) When the situation is such that the Reynolds numbers both for the wall and the packing are above the critical value and hence both the packing and the wall are thoroughly covered by the liquid. For the parameter C then there results

$$C = \frac{P_{\mathrm{p}}}{P_{\mathrm{w}}} \sin \alpha \tag{4-43}$$

and we can see that the parameter C is solely a function of the geometry of the packing and the column (wall) which is in agreement with the tenets of the model leading to the boundary condition Eq.(4-25). In addition, substituting corresponding expressions for the quantities P_{p} and P_{w} using geometry considerations we find that the parameter C is directly proportional to the ratio of the column to particle diameter, d_c/d_p, which is in agreement with experimental finding. Direct proportionality between C and d_c/d_p was also derived theoretically by Jameson [177,178].

Cases b) and c) representing better wettability of the wall and packing respectively and low irrigation rates predict C to increase linearly with the mean irrigation rate (case b) or inversely with the mean irrigation rate (case c).

Since the wettability of the two surfaces, characterized by the appropriate contact angles, may vary from place to place as well as with time, the first three cases (a,b,c) may qualitatively explain the complexities of the behaviour of the columns at very low irrigation rates. Admittedly, any dependence of C on the liquid load is at odds with the premises of the diffusion model of liquid distribution.

Brignole et al. [189] also objected that the dimensionless equilibrium wall flow depends on the liquid rate and proposed a boundary condition expressing the dimensionless wall flow as a quadratic function of the ratio d_c/d_p with a single liquid load dependent parameter. They found, however, that dependence of the parameter for Raschig rings on the liquid load, is very weak.

Furzer [203] interpreted in an interesting way the boundary condition (4-24) in terms of the transfer coefficients and efficiencies for the transfer of liquid to and from the wall. From experiments with 25 mm Lessing rings in a 600 mm diameter column he found these two efficiencies to be approximately the same and equal to 0.012 and 0.013. These low values provide a proof that at least Lessing rings are poor at transferring the liquid to and from the wall of a packed column.

Numerical values of C

Determination of the parameter C mostly requires the experimental measurement of the magnitude of the wall flow. This can be accomplished by experimental techniques separating carefully at the bottom of the bed the wall flow from the flow through the packing in the vicinity of the wall. This task is not very easy. For various constructions of the device separating the wall flow see e.g. Porter [175], Jameson [178], Porter and Templeman [57], Dutkai and Ruckenstein [55], Onda et al.[54] Brignole [189], Farid and Gunn [193] and Staněk and Kolář [4].

From experimental data it seems well established that the parameter C (i.e. the ratio of the dimensionless density of irrigation to the dimensionless wall flow at equilibrium) is directly proportional to the ratio of the column to particle diameter d_c/d_p. Staněk and Kolář [13] correlated their data as well as the data of Dutkai and Ruckenstein [55], Zarzycki [56] and Porter and Templeman [57] and obtained the following correlation

$$C = k \frac{d_c}{d_p} \tag{4-44}$$

in which the proportionality coefficient equals

$$k = 0.181 \quad \text{for Raschig rings}$$

and

$$k = 0.365 \quad \text{for spheres.}$$

From the above values of the constant k it is thus seen that the effect of the walls is stronger for the bed of Raschig rings.

The value of $C = 1.7$ computed from the data of Porter and Jones [173] for 1/2 inch Raschig rings in a 6 inch column fits also reasonably into the above correlation with a deviation toward stronger effect of the wall flow. Also the values of C computed from the data of Templeman and Porter [175] for 1/2 inch Raschig rings in columns 2,3,4 and 6 inch in diameter fall well into the above correlation.

For Pall rings Hoek et al.[206] report $k = 0.36$. The constant k of the correlation (4-44) evaluated from the data of Hanika et al.[182] presented in the form of a plot of W versus d_c/d_p for three types of porous catalyst pellets yielded $k = 0.392$ i.e. a value close to that for Raschig rings.

Finite values of the parameter C are a consequence of structural inhomogeneities of the bed near the confining walls. However, there is no direct relation [4] between the extent of the wall flow and increased void fraction in the wall region. The conditions are thus more complicated and the different curvatures of the wall and the packing also probably play an important role.

Fig (4-3) showed the contours of constant density of irrigation for a uniformly irrigated bed for $C = 2$ and $B = 6.7$. Using the last correlation for the parameter C we find that the depicted situation would occur in a column with d_c/d_p equal 5.5 packed with spheres, while for a packing of Raschig rings this situation would occur at d_c/d_p equal to 11.

For the Raschig rings this ratio is above the usually recommended [174] value of d_c/d_p equal to 10 and yet the observed wall flow effects is strong. Some authors, e.g. Baker et al. [166] recommended an even smaller critical d_c/d_p ratio of 8 where the wall flow would be even stronger. Accordingly, the recommendation of Jameson [178] that this ratio be at least 25 is more realistic.

Numerical values of B

The values of B were determined experimentally by Staněk and Kolář [13] for a bed of spheres 15 and 20 mm in diameter with no appreciable effect of particle size and the mean density of irrigation with the result

$$B = 6.7 \quad \text{for spheres.}$$

The same work yielded for Raschig rings 15 and 25 mm in diameter also values essentially independent of particle size

$$B = 7.0 \quad \text{for Raschig rings.}$$

Here, however, a certain trend has been detected for B to decrease slightly at low density of irrigation.

The data of Dutkai and Ruckenstein [55] for Raschig rings 10, 15 and 25 mm in diameter, processed according to the formulae (4-38) and (4-39) yielded slightly lower values between 6.95 (for smaller rings) and 4.24 (for larger rings). The dependence on d_p, however, seems to come entirely from the data on the spreading coefficient used in the calculation. An important conclusion following from these experimental measurements in two columns 150 and 250 mm in diameter is that there is no detectable effect of d_c on B.

Dutkai and Ruckenstein [55] data for 15 mm Intalox saddles yield $B = 7.8$ or 8.2 for the 150 and 250 mm column respectively.

Prchlík et al. [185] published a single value of $B = 1.5$ found for a bed of nickel on kieselguhr catalyst pellets for which C equalled 13.3.

The above data for B and C provided a surprisingly good agreement [12] between the experimental and theoretical density of irrigation profiles for the case of initial irrigation of the top of the bed by a "point source" and a disc distributor ($r_1 = 1/2$) for the heights of the bed in which non-negligible amounts of the wall flow were apparent.

As far as the effect of the physical properties of liquid on the parameters B and C is concerned very little is known. From the work of Staněk and Kolář [4] and Onda et al. [54] it can be concluded that C is slightly higher if the surface tension of the irrigating liquid is reduced from the value typical for water. B, on the contrary, is decreased [4] by decreasing the surface tension and by increasing the viscosity of liquid. These conclusions were made [4] on the basis of experiments with water, glycerol, aqueous solution of detergent and a 7% aqueous solution of isobutanol.

4.3. THE POTENTIAL FLOW MODEL

Gunn [194] argued that the "diffusion" model in Eq.(4- 5) is incomplete as the potential for the radial spreading of liquid is only the axial component of superficial velocity. In reality, the underlying stochastic process involves also axial dispersion whose potential is the radial liquid velocity component. This lead Gunn to the formulation of the potential, p, comprising both contributions. This potential may, in addition to the axial, V_z (V_z being identical with f), and radial, V_r, velocity components include pressure and gravitational forces.

The potential obeys the following differential equation

$$\frac{\partial^2 p}{\partial z^2} + \frac{1}{r}\frac{\partial}{\partial r}\left(r\,\frac{\partial p}{\partial r}\right) = 0 \tag{4-45}$$

with the velocity components being related through the permeability of the packing, K_p, (an analog of D) to the potential by

$$V_z = -K_p \frac{\partial p}{\partial z} \tag{4-46}$$

$$V_r = -K_p \frac{\partial p}{\partial r} \tag{4-47}$$

The bulk packed region extends up to a certain radius while the remaining annular region is the wall region characterized by a different permeability K_w.

Gunn [194] presented analytical solutions of the potential model in the form of the stream function in terms of the Bessel functions with the parameter being the ratio of permeabilities of the bulk packing and the wall zone. The width of the wall zone region was taken to be one particle diameter wide. The ratio of permeabilities is thus a characteristic of the tendency of the column to form the wall flow.

The liquid that leaves the bulk of the packing enters the wall region so that the material balance is preserved. Analytical solution of this model gives qualitatively the same picture as the solution of the diffusion model.

The principal advantage of the potential model is that the potential may also incorporate more complex phenomena such as e.g. phase interactions at two phase flows, etc. A substantial difference from the diffusion model is that the wall region has a non-zero width.

Experimental investigations of the extent of the wall flow and the influence of the gas velocity on its magnitude on the basis of the potential model were made by Farid and Gunn [193]. The experimental results have shown the model to predict well liquid distribution at gas velocities reaching at least 50% of the flooding rate. Only in this case were the ratios of permeabilities obtained found to be independent of the initial distribution of liquid, which is a prerequisite for the validity of the model.

There is a definite disadvantage of the potential model in that the relative extent of the wall region varies with particle size as does the ratio of permeabilities.

For 1 and 1.5 inch Raschig rings the permeability of the wall region was found to increase with increasing gas velocity but these results were affected by the extent of the model wall region.

Furzer [202], however, experimented on a 600 mm in diameter column packed with Lessing rings and reported major deviations from Gunn's potential model. The significant deviations were in the wall region where experimental changes of the density of irrigation with the depth were much slower than predicted by the potential model.

4.4. THE EFFECT OF FLOW MALDISTRIBUTION ON MASS TRANSFER

The adverse effect of flow distribution on mass transfer has two aspects. One is the assessment of the loss of separating efficiency of the column due to maldistribution. The other is the effect of flow distribution on the design and scale-up of the columns. The latter aspect was intensively studied by Billet [236,237]. The effect of flow maldistribution in distillation columns packed with high performance packings was reviewed by Olujic and de Graauw [336].

The former aspect is basically the problem of quantification of the extent of flow maldistribution and relating it to the transport phenomena taking place in the column.

As already mentioned there are two types of flow maldistribution: large scale and small scale maldistribution.

Small scale maldistribution

On the small scale it is acknowledged that especially at low liquid rates the liquid, as it descends down the bed, follows preferential paths. Preferential paths have already been observed by Lespinasse and Le Goff [231]. Charpentier et al. [232], in addition, observed flow in the form of drops. As the liquid load in the column increases the flow tends to create more and more film over the surface of the particles and less and less drops and rivulets.

Refre and Hellincx [190] observed, using a conductivity probe to measure the local flow conditions, that the spreading of liquid maintains the thickness of the liquid film in the bed at a constant value until the bed is completely wetted. Only then does the film thickness increase. Their measurement pictured the flow in the packing as a combination of the preferential path and the rivulet concept. The situation, however, was not found to be static. It rather changed from one path to another as well as from an instant to the next with the local streams varying very significantly.

The conclusions of measurements of Refre and Hellincx [190] clearly contradict the earlier concept of steady preferred paths from which the liquid cannot escape. Groenhof [196], for instance, reported that the preferred path do not vary significantly with time, their pattern is not changed in a statistical sense by flooding the column before operation. The pattern, however, changes completely after repacking the column.

These small scale maldistribution phenomena have often been referred to by the not quite well defined notion of "channelling".

Hoek et al.[206] in their experimental study of liquid distribution in a 500 mm in diameter column using small catch pots and various packing overcame the ambiguity of the notion of channeling. They defined the channel as a location within the packing where the irrigation rate is 50% higher than the average and this event occurs over an area equivalent to at least two particles. The experimental data of Hoek et al.[206] evaluated as the maldistribution factor, Mf^2, defined essentially as the square of the relative standard deviation Mf in Eq.(4-15), indicated that the maldistribution factor Mf^2 as well as the number of channels per square meter of column cross sectional area vary with the size of the catch pot relatively to the size of the packing. It was found that the maldistribution factor, Mf^2, measured at the particle scale, i.e. with the area of the catch pot equalling d_p^2 is constant and in the range of particle size between 10 and 50 mm investigated experimentally Mf^2 will be equal to 0.5.

Also the number of channels, as defined above by Hoek et al.[206], will vary with the size of the catch pot. If, however, the evaluations are made again at the pellet scale the number of particles per channel in a given cross section will be constant and range between 30 and 40 particles.

The degree of maldistribution was found [206] always to be higher when the mean irrigation densities dropped to 0.005 ms^{-1} and Mf^2 aproximately doubled.

The degree of maldistribution given by $Mf^2 = 0.5$ on the pellet scale and when the number of particles in a column cross section per channel is 30-40 defines a relatively stable state of irrigation (excluding the large scale maldistribution effects) that is termed by Hoek et al. [206] "natural flow". A distributing device that should bring about a more uniform flow in the bed, thanks to its greater number of pour points (rivets, jets, nozzles) per unit column cross section, expressed in terms of particle cross sectional area will, however, succeed to do so for only a shallow section of the bed under the distributor. At only one layer of particles deep below this distributor the actual distribution is considerably impaired and in about ten layers of particles the flow again reaches the "natural" conditions, given by the above value of the Mf^2 on the particle scale. If, on the other hand, the number of pour points of the distributor is smaller than the number of channels corresponding to the "natural" flow for the given packing, it will take several layers of packing to establish the "natural flow" conditions. This process, however, is much slower and takes a greater depth of the bed. A practical conclusion is that the number of pour (or drip) points of the distributor should correspond to the natural density of channels found by Hoek. In large diameter columns the number of pour points should be about 60 to 100 per square meter.

Small scale maldistribution or channelling will be responsible for impaired efficiency of mass transfer due to the reduced local interfacial area, variable local liquid hold-up and variable local liquid velocity.

The last factor in turn will cause locally variable ratio of the gas and liquid velocity, i.e. the slope of the operating line on the Mac Cabe-Thiele diagram. The driving forces for interfacial transfer are thus reduced both locally and in the column as a whole. The most serious effect of impaired column efficiency due to the maldistribution will thus occur in columns with a high number of transfer units corresponding to the ideal uniform flow and in separations where the composition of phases is near equilibrium, see e.g. Schlünder [210].

Groenhof [196] concludes that the change of the slope of the operating line due to the maldistribution appears to be mainly due to local changes of liquid velocity because the flow rate variations of the gas phase are evened out faster by the action of the pressure drop.

Extensive experiments of Groenhof [196] in columns 0.14, 0.29 and 0.44 m in diameter packed with 8.4 mm Raschig rings to 2 m height at relatively low liquid load (0.825 and 0.50 $kg\,m^{-2}s^{-1}$) aimed at measuring the flow maldistribution, local residence time distribution and local concentration during stripping in an acetone-water-air system, yielded interesting results. Groenhof defined maldistribution indices for the local velocity, local residence time and local concentration all in the form of standard deviations of these quantities (analogously to Eq.(4-15) for *Mf*) normalized by the mean value of the quantity examined.

His data [196] showed extensive maldistribution with the flow maldistribution index ranging between 0.39 and 0.51. These values almost did not change with the column diameter and showed also good reproducibility after repacking the column. This clearly indicates that although the pattern of the flow paths changes strongly, the degree of maldistribution does not.

The maldistribution index for the residence time at different locations on the bottom column cross section gave values at least ten times smaller (0.04) than corresponding flow maldistribution indices. From this finding one can conclude that probably due to the local variations along the depth of the bed individual liquid paths as a whole differ little one from another (note that the residence time is a property of the path as a whole while the outlet liquid velocities are characteristic only of the short section of the bed just above the bottom due to the radial mixing). This can be further supported by arguing that the residence time is indirectly proportional to the irrigation rate and directly proportional to the hold up. Since the hold up is approximately proportional to the (2/3)-th power of the irrigation rate the residence time is proportional only to the (-1/3)-th power.

Using this argument Hoek et al.[206] derived that for $Mf^2 = 0.5$ on the particle scale the corresponding Mf^2 for the residence time should equal 0.08.

The maldistribution index, *Mf*, for the local effluent concentrations of acetone was found by Groenhof [196] to be 0.17 which is less than one half of the flow maldistribution index of 0.39 for corresponding flow conditions. The explanation is as before: The local outlet concentration results from mass transfer along the whole flow path in contrast to the local flow rate. The fact that the maldistribution index for the concentration is higher than that for the residence time may be explained by arguing that the effect of local variations of liquid velocity on the residence time tends to be smoothed out by changes of the holdup while corresponding changes of the interfacial area, important for mass transfer are not as directly related to the flow rate. In addition, Nakamura et el.[205] found experimentally the radial distribution of the liquid hold up to be almost uniform.

Kennedy and Jaffe [240] also observed strong liquid flow maldistribution in a commercial-scale trickle bed reactor for desulfuration by measuring the residence time distribution. The associated maldistribution was well modelled by two parallel-flow reactors. The temperature of the reactor would have to be increased by 80 K in order to compensate for the impaired function of the reactor due to maldistribution. The residence time measurement, however, do not give insight into the mechanism of the flow maldistribution.

Large scale maldistribution effects

Large scale maldistribution appears due to either nonuniform initial distribution and/or the wall flow effects.

Porter and Jones [175] demonstrated the effect of the wall flow on the efficiency of a distillation column 6 inch in diameter packed to a depth of 5 feet by 1/2 inch Raschig rings for the distillation of a water/methanol mixture. They measured the concentration of methanol in the packing and the wall flow stream at the bottom of the column. The magnitude of the wall flow, measured in this study exceeded slightly 50% of the total flow independently of the boil-up rate. This magnitude of the wall flow is considerably higher than the 31.5 % which would follow for the given d_c/d_p from the correlation (4-44). The difference can be explained by the fact that the device employed to separate the wall flow separated also part of the flow over the packing in a 1/8 inch wide annulus adhering to the wall.

In all experiments [175] the concentration of methanol was substantially higher in the wall flow than in the effluent stream from the packing. This clearly indicates the unfavourable conditions for vapour/liquid interfacial transfer in the wall region.

In addition, the concentration difference between the packing and the wall flow increased with decreasing boil-up rate. The increase, however, was mainly due to the decrease of the methanol concentration in the stream from the packing. Understandably, the decreased boil-up rate improved the transfer conditions on the packing but not in the wall flow.

Also Groenhof [196] experiments with acetone stripping showed the acetone concentration in the liquid near the wall to be always higher which demonstrates clearly the adverse effect of the wall flow on mass transfer.

The number of pour points of the distributor and their pitch also affect, for geometry reasons, the uniformity of irrigation in the wall region. This quality of the flow uniformity in this case is termed "the degree of coverage" and proper irrigation of the peripheral regions of the column is found to have profound effects on the separating efficiency [216].

In general, packed columns have distinct advantages over plate columns such as for example low hold-up. Their use as large scale columns, however, has been hampered mainly by design uncertainties due to the effect of maldistribution. Small scale maldistribution, given by the "natural" flow, although quite severe in character, has only a minor effect on the separation efficiency. This is very different from large scale maldistribution that can cause severe losses in the separation efficiency as a whole [216]. Nevertheless, increased understanding of flow maldistribution phenomena in packed columns have sparked what is called a "quiet revolution" [336] in changing over from tray to packed columns.

Mathematical models

Huber and Hiltbrunner [176] developed three models quantitatively describing the effect of the liquid flow maldistribution on the separating efficiency of a column. They further developed qualitative concepts of Manning and Cannon [233] and Mullin [234].

The first model utilizing Mullin's [234] concept views the column as consisting of two parallel sections with no lateral mixing between them. At the bottom the gas (vapour) is evenly distributed into the two sections while at the top both outlet gas streams are mixed condensed and returned into the two sections in proportions given by the maldistribution factor, ℓ, taken as a fraction of the liquid rate under ideal uniform conditions. Under the effect of liquid flow maldistribution the liquid flow in one section is by the fraction ℓ (i.e. $f_0\,(1+\ell)$) in the other section by the fraction ℓ lower (i.e. $f_0\,(1-\ell)$).

The balance on the transported species was formulated under standard assumptions plus the simplification that the relative volatility is not excessively large. A plot of the number of transfer units (NTU) as a function of the maldistribution factor, ℓ, in the log-log coordinates computed on the basis of this model indicates that the detrimental effect of the liquid flow maldistribution is the higher the larger the number of transfer units of the column under the uniform flow conditions. It can be shown that for a column operated under the total reflux the reflux in both section of the column will be finite.

Under simplified conditions one can write for a column with a large (infinite) NTU_{id} under ideal flow conditions that NTU will reduce to

$$NTU = \frac{2-\ell}{\ell} \quad \text{if} \quad \ell > (\alpha - 1) < 0.1 \tag{4-48}$$

where α is the relative volatility.

The second model of Huber and Hiltbrunner [176] is based on the previous one but further proposes that the column is divided along its height into N_{S} sections and that a fraction, m', of each stream in the two parallel sections is mixed together and returned back.

The mathematical model incorporating this type of cross mixing was integrated numerically and the results were presented in the form of a plot of NTU as a function of cross mixing in the log-log coordinates with the degree of maldistribution, ℓ, as a parameter.

The cross mixing, which was attributed by the authors to the action of the gas flow, was characterized by the product $N_{\mathrm{S}}m'$. The results shows that with increasing $N_{\mathrm{S}}m'$ the number of transfer units increases.

Expectedly, lateral, or cross flow mixing counters the effect of liquid flow maldistribution and improves the column separating efficiency.

It was interesting to find that the results were almost the same for different values of N_S provided that the product $N_S m'$ was the same. From the standpoint of improved separating efficiency of the column it appears to be more efficient to mix smaller fractions of the liquid stream more frequently (greater N_S) than to mix greater fractions of liquid (greater m') less frequently. From the published results of Huber and Hiltbrunner [176] it can be seen that a significant increase of NTU beginns when $N_S m'$ assumes value of about 0.2.

The third model [176] considers continuous cross mixing of liquid along the column length. A simplified formula has been given to calculate the number of transfer units as a fraction of the this number under ideal conditions, NTU_{id}.

$$\frac{NTU}{NTU_{id}} = \frac{1}{1 + \frac{f}{f_G}\frac{8c}{m}} \tag{4-49}$$

where f_G is the superficial gas velocity and m is the fractional cross mixing of liquid per transfer unit. The parameter c is related to the reflux ratio, R_F, and the fractional maldistribution, ℓ, by

$$c = \frac{1}{60}\left(5\frac{\ell}{R_F} + 2\ell^2\right) \tag{4-50}$$

Stichlmair and Stemmer [209] have also used the concept of the bed consisting of a large number of hypothetic parallel not communicating sections with different liquid loads and applied it to the case of the counter-current flow of hot water and air in a packed bed column. The process within the bed was monitored by a large number of thermocouples measuring the temperature of water at various radial and axial conditions permitting construction of the isotherm pattern existing within the bed. In case of uniform liquid and gas flow the isotherms should be, of course, horizontal lines. Experimental results, however, exhibited severe deviations of isotherms from horizontality.

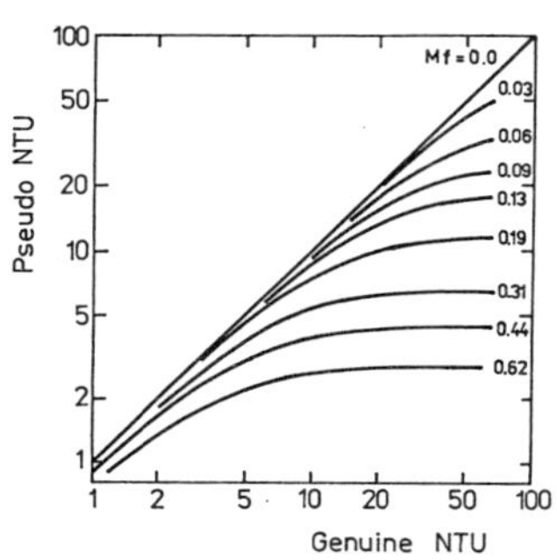

Fig.4-5
Correlation [209] between genuine and pseudo NTU. The degree of maldistribution is expressed by Mf, Eq.(4-15).

Stichlmair and Stemmer [209] constructed a plot of the real number of transfer (pseudo) units versus the ideal transfer (genuine) units corresponding to the plug flow (see Fig.4-5) with the liquid flow maldistribution fact, M_f, as a parameter. This factor was again defined as the standard deviation of the normalized liquid velocity in individual sections of the column, see Eq.(4-15). Similarly, Huber and Hiltbrunner [176] plot the number of real transfer units which progressively decreases with increasing number of ideal number of transfer units and ultimately reaches a plateau a where the real number of transfer units no longer increases with the depth of the bed.

A disadvantage of such plots is that the result depends on the distribution of liquid between the hypothetical parallel sections composing the column. Stichlmair and Stemmer therefore constructed another plot of again the pseudo versus genuine transfer units with the concentration maldistribution factor as a parameter. The concentration maldistribution factor was computed from concentrations (in this case replaced by temperatures) as a standard deviation of the normalized values on several horizontal levels in each of the parallel section. Corresponding lines in the number of genuine versus the number of pseudo transfer units plot for individual concentration maldistributions are nearly parallel with the diagonal. One can now consider various types of maldistribution varying along the height of the column. A distinct disadvantage for practical use, however, is that the characteristic of maldistribution has to be evaluated from mass transfer experiments; mere flow maldistribution data do not suffice.

None of the above models incorporated the effect of gas flow maldistribution or axially variable liquid flow distribution. Nevertheless, the practical conclusions that can be inferred are very instructive for practical use.

In columns with the ratio d_c/d_p about 10 the cross mixing is sufficiently intensive so that only with extremely strong liquid flows maldistribution would cause a significant loss of efficiency. If, however, this ratio exceeds 30, as is the case in industrial columns, lateral mixing may not be intensive enough to offset the influence of liquid flow maldistribution. All that, of course, concerns small scale maldistribution. The problem of liquid flow maldistribution will also increase considerably with the height of the column even if the degree of maldistribution remains the same. Increased height means a greater number of transfer units and therefore a stronger detrimental effect of the liquid flow maldistribution.

The undesirable effect of small scale maldistribution is suppressed, as has been seen, by radial mixing. However, large scale maldistribution, if it develops along the wall of the packed column, cannot be easily compensated by radial mixing. This can be a problem in large diameter columns even if the wall flow in this case is not severely large.

Porter and Jones [215] conclude on the basis of observation that D and $HETP$ are proportional to d_p and gas flow pattern is similar in geometrically similar columns so that the scale up of columns can be pased on experiments on geometrically similar columns. The results, however, will be more accurate for commercial size (d_p greater than 10 mm) packings.

Kunesh et al. [170,214] ran an extensive controlled maldistribution program. The experimental part was distillation of cyclohexane/normal heptane in a 1.2 m in diameter column equipped with an adjustable liquid distributor. This device permitted control of the initial distribution of liquid and hence to change the distribution patterns without having to change the distributor. Their experimental results were interpreted in terms of HETP.

At total reflux and usual assumptions leading to the Fenske equation

$$NTP = \frac{h}{HETP} = \frac{1}{\ln\alpha} \ln\left[\left(\frac{X}{(1 - X)}\right)_{\text{top}} \left(\frac{(1 - X)}{X}\right)_{\text{bottom}}\right] \tag{4-51}$$

a plot of $\ln(X/(1 - X))$ versus bed depth where X is the mole fraction of the component should yield a straight line. The slope of this straight line should be $[\ln(\alpha)/HETP]$ where alpha is the relative volatility considered to be a constant equally as $HETP$.

A curved line from this plot with the slope becoming steeper from the top down indicates that the ratio of liquid to vapour flow is changing as the spreading action of the packing corrects the initial maldistribution and establishes "natural" flow within the bed. This will persist unless the wall flow effects flatten the slope again.

Apart from this picture, however, more complicated courses on the above plot were observed indicating no recovery in the system at all.

More complicated behaviour of the effect on $HETP$ could be explained by applying the Zone-stage model of Zuiderweg et al. [236,216]. This model uses the concept of "basic" or "inherent" $HETP$ which is a function of the packing and system only. The bed is divided axially into the appropriate number of theoretical stages (height equal to basic $HETP$) and radially into a number of concentric zones, recommended to be 2-3 times the particle diameter. The outermost zone includes the wall flow. Integration is carried out numerically for each zone with averaging for each zone-stage combination. Iterative approach is needed until appropriate end conditions are achieved.

The Zone-stage model seems to match the data very well and confirms the idea of basic $HETP$ independent of initial distribution. This opens up a good possibility of calculating the effects of varying degrees of maldistribution.

As shown by Zuiderweg [216] the maldistribution effect on the separation efficiency (e.g. as a ratio of effective and basic transfer trays) can be expressed in terms of the Fourier number, Fo

$$Fo = \frac{(d_p/2)}{D \cdot (HETP)} \tag{4-52}$$

for other conditions constant.

Kunesh et al.[170,214] found that columns have a surprising tolerance in their separation efficiency to moderate continuous changes of the initial irrigation rate (e.g. from the center to the wall and vice versa). The separating efficiency, however, is severely reduced if there are discontinuities in the initial profile of the irrigation rate (e.g. tilt of the distributor leaving part of the top surface of the bed dry).

Gas flow maldistribution effects on the gas-side mass transfer coefficient

The role of gas flow maldistribution in affecting mass transfer processes in irrigated packed towers was investigated by Krebs [242, 243]. The effect of gas flow maldistribution on mass transfer is expected to be potentially strong in gas-side controlled processes. Krebs investigated evaporation of isopropanol--water mixture into dry air in a bed of 15 mm Raschig rings. In contrast to evaporation of a given solute into various gases, or, on the contrary, evaporation of various solutes into the same gas, this method yields the ratio of gas-side mass transfer coefficients for isopropanol and water. The ratio of mass transfer coefficients is obtained in the form of the ratio of the number of transfer units, *NTU*, for the two solutes.

Krebs [242, 243] found the dependence of the ratio of *NTU* for isopropanol and water to be much weaker in packed column than in wetted wall column experiments. The weaker dependence was explained by the nonuniformity of the gas flow in the packed bed.

If the bed was modelled as a series of channels of two different diameters, the weaker dependence could be attributed to the split ratio of the gas flow between the larger and the smaller channels. This approach is, in effect, equivalent to the Martin's two-zone model [88] with the larger diameter channels corresponding to the wall region with increased void fraction. Corresponding optimum values of the split ratio indicated that about 90 to 95 % of the gas flow takes place through the larger channels. This would indicate an extremely strong peripheral flow of gas.

4.4.1. The wall flow "wipers" and liquid flow redistributors

The detrimental effect of the wall flow was recognized as early as 1931 by Berl [228] who attempted to counter it by developing a saddle packing with lesser tendency to form the wall flow. Also Kirschbaum investigated this effect in the same year [229] and in 1934 patented a redistributor [230] restoring the uniform distribution in the column.

The question arising in connection with flow redistributors is the proper distance, or spacing, between individual redistributors. Prchlík et al.[186] recommend on the basis of experiment with Raschig rings, glass spheres and Intalox saddles and with d_c/d_p ranging between 4 and 16.7 the following expression for the dimensional distance, z_r, to locate a distributor when the initial distributor is a "point source"

$$z_r = 0.042 \frac{d_c^2}{d_p} - \frac{d_c}{0.064} \ln\left(\frac{W(\infty) - W}{W(\infty)}\right) \tag{4-53}$$

This distance will keep the wall flow below the limit W.

The function of the wall flow deflecting rings or wall flow wipers in general is to improve the ability of the wall as a vertical surface to return the liquid back into the packing. Kolev et al. [200] showed that the width of the deflecting ring need not exceed the size of the packing element.

Deflecting rings or wall "wipers" rectify the conditions mainly in the peripheral region of the column. Zuiderweg et al. [216] found that the efficiency of the packed column is very sensitive to initial distribution of liquid and a particular care should be taken that the peripheral regions are irrigated adequately.

With small packings in large columns, or in deep beds redistribution is necessary. It is better if the redistributor redistributes both liquid and gas (vapour).

Staněk et al. [221] investigated the function of the deflecting ring (wall "wiper") both experimentally and theoretically on the basis of the "diffusion" model. The deflecting ring was considered to be a ring of dimensionless radius r_1 extending from this radius up to the column wall ($r = 1$).

The ability of the diffusion model to predict the function of such a ring was tested first on a single ring placed directly under the distributor which would normally distribute the liquid uniformly over the top surface of the packing. What happens under such circumstances is that the uniform initial distribution is changed and can be described by the following equations

$$f = 1 \quad \text{for} \quad z = 0 \quad \text{and} \quad 0 \leqq r < r_1 \tag{4-54}$$

$$f = \infty \quad \text{for} \quad z = 0 \quad \text{and} \quad r = r_1 \tag{4-55}$$

$$f = 0 \quad \text{for} \quad z = 0 \quad \text{and} \quad r_1 < r \leqq 1 \tag{4-56}$$

This, in effect, means that within the radius r_1 the top surface is irrigated uniformly while all liquid, that was to be fed originally between the radius r_1 and the wall, leaves the deflecting ring exactly on its inner edge (radius r_1) while the remaining annulus remains dry.

Corresponding solution of the diffusion equation (4-5) for the above initial conditions and the boundary condition (4-25) can be obtained most easily from the combination of the solutions (see Table 4-1) for the disc distributor of radius

r_1 and the circular distributor, both being properly weighted in order to preserve the mass balance on liquid. The result is [221]

$$f = \frac{C}{1 + C} + \sum_n \left\{ \frac{[(q_n^2/B - 2C]^2 J_0(q_n r) \exp[-q_n^2 z H]}{\{[(q_n^2/B - 2c]^2 + q_n^2 + 4C\} J_0^2(q_n)} \cdot \left[\frac{2r_1^2 J_1(q_n r_1)}{q_n} + (1 - r_1^2) J_0(q_n r_1) \right] \right\} \tag{4-57}$$

Experimental results [221] to check the validity of the above expression were carried out in a 189 mm diameter column packed with 25 mm Raschig rings. The wall "wipers" used were 20 and 30 mm wide and at the outer edge of the bottom face provided with a notch meant to facilitate the draining of the liquid back into the packing. The experiments showed a reasonably good agreement.

A more interesting modification of this experiment, however, was investigated by Semkov et al. [222] when several deflecting rings were placed evenly spaced at some distance below the distributor. The theoretical part of the problem was solved as follows:

The initial distribution is given by the type of the liquid distributor at the top of the bed. Solution of the diffusion equation (4-5) for the depth between the top and the location of the first deflecting ring can be found either in the Table 4-1 or from Eq. (4-29). This solution is an initial condition for the packed section between the first and the second deflecting ring but only up to the radius r_1. For the remaining annular space the two conditions (4-55) and (4-56) apply.

Solution of the diffusion equation for the packed section between the second and the third deflecting ring is obtained relatively easily with the initial condition described above as all the solutions involved are in the form of a series of the Bessel functions.

Analogously the distribution of liquid just above the third reflecting ring is the basis for the formulation of the initial condition for the section between the third and the fourth deflecting ring in region up to the radius r_1 while for the remaining annulus the equations (4-55) and (4-56) apply.

One thus obtains a recurrent formula in terms of the Bessel functions for the distribution of liquid between the k-th and the $(k + 1)$ - th deflecting ring. The formula can be found in the original work [222]. It is obvious, however, that the solution will not differ very much when k is sufficiently large as we reach the quasi steady state periodic solution.

Experimental measurements carried out by Semkov et al. [222] with 20 mm wide deflecting rings using a different number of the rings (2 or 3) and different spacings (100 or 200 mm) did not agree well with predicted liquid distributions. The deviations in the wall region were always such that the real flow near the wall was less than predicted by the theory as if the effective width of the deflecting rings was greater than the actual size.

An optimization routine which also took into account the width of the deflecting rings as an adjustable parameter showed the optimum size to be approximately equal the size of the packing element. rThis led the authors to the speculation that once the deflecting rings are located within the packing the liquid does not drain from the deflecting rings directly from the inner edge. Instead, the draining is mediated by the particles (in this case Raschig rings) in contact with the deflecting rings. This effectively decreases the radius r to which all liquid from the annular region near the wall is deflected. This speculation was further confirmed by visual observation through the walls of the perspex glass column. This phenomenon was further investigated on the basis of statistical theory by the same authors [223].

The above results of Semkov et al. [224] were further used to optimize the spacing of the wall flow deflecting rings. The optimum distance in terms of particle diameter was found to depend little on the column to particle diameter ratio, d_c/d_p, and was approximately directly proportional to the width of the deflecting ring expressed as a fraction of the particle diameter.

4.4.2. A mathematical model of the effect of flow distribution on absorption of a poorly soluble gas

Absorption of a poorly soluble gas will be controlled largely by the resistance on the side of the liquid film and therefore mostly by the maldistribution of the liquid flow. This opens up the possibility of utilizing the diffusion model of liquid distribution, including the existence of the wall flow, to formulate a model of absorption in the presence of nonuniform distribution of liquid and the presence of the wall flow [239].

Let us consider a bed h deep and designate the concentration of the absorbate by c. We have already noted that the principal mechanism of radial mixing under trickle flow is by radial spread of the liquid proper. Accordingly, we can write the balance for the absorbing medium in the liquid on the packing in a manner analogous to Eq. (4-12) for the spread of liquid tracer. In addition, one has to account for the interfacial transfer. Thus we obtain

$$\frac{\partial(fc)}{\partial z} - H\left[\frac{\partial^2(fc)}{\partial r^2} + \frac{1}{r}\frac{\partial(fc)}{\partial r}\right] - N_p(c - c^*) = 0 \tag{4-58}$$

where N_p designates the number of transfer units defined in case of a poorly soluble gas on the basis of the liquid side volume mass transfer coefficient $k_L a$

$$N_p = \frac{k_L a h}{f_0} \tag{4-59}$$

c^* designates concentration of the absorbing species at the interface. It is taken to be at equilibrium with the gas phase composition and can be therefore computed from the gas phase concentration and corresponding Henry constant.

Eq. (4-58) can be arranged by performing the differentiation and after simplification using the balance on liquid in Eq. (4-5) we obtain

$$f\frac{\partial c}{\partial z} - H\left[f\frac{\partial^2 c}{\partial r^2} + 2\frac{\partial f}{\partial f}\frac{\partial c}{\partial r} + \frac{1}{r}f\frac{\partial c}{\partial r}\right] - N_p(c - c^*) = 0 \tag{4-60}$$

The boundary condition characterizing the transfer near the wall is also written in the form of a balance

$$-\frac{\partial(fc)}{\partial r} = B(fc - CWc_w) \quad \text{for } r = 1 \tag{4-61}$$

where the left hand side term expresses the net dimensionless flux of species into the wall flow, expressed, on the right hand side, as a difference of the fluxes into the wall flow (the first term) and back into the flow over the packing surface (the second term). The concentration of the absorbate in the wall flow, of course, is different and was therefore designated by the symbol c_w. We note that c_w is a function of the axial coordinate only, while c, the concentration within the packing, is generally a function of both the radial and axial coordinate (for axisymmetric flows).

One can again [239] perform the differentiation and simplify the result with the aid of the boundary condition for the flow of liquid in Eq. (4-25). Thus we have

$$-f\frac{\partial c}{\partial r} = BCW(c - c_w) \quad \text{for } r = 1 \tag{4-62}$$

The above equations contain the so far unknown concentration of the absorbate in the wall flow. This can be obtained from the mass balance on the species in the wall flow in the form

$$\frac{\mathrm{d}(Wc_w)}{\mathrm{d}z} = -2H\frac{\partial(fc)}{\partial r} + N_w(c_w - c^*) \tag{4-63}$$

where N_w designates the number of transfer units in the wall flow given by

$$N_w = \frac{2k_{LW}h}{f_0 R} \tag{4-64}$$

We note that k_{LW} is the liquid side mass transfer coefficient referring to the conditions on the wall and $(2/R)$ is the specific surface of the wall per unit volume of the bed. The derivative in the first term on the right hand side of the balance on the absorbate in the wall flow is, of course, taken in the point $r = 1$ as we work now with an ordinary differential equation.

Eq. (4-63) can be further rearranged by substituting from the boundary condition for the liquid (4-25) with the result

$$\frac{d(Wc_w)}{dz} = 2HBCW(c - c_w) + c\frac{dW}{dz} + N_w(c_w - c^*) = 0 \tag{4-65}$$

By performing the differentiation on the left hand side and after some manipulation we obtain

$$W\frac{dc_w}{dz} + (c_w - c)\frac{dW}{dz} = 2HBCW(c - c_w) + N_w(c_w - c^*) \tag{4-66}$$

The last form of the balance on the absorbate in the wall flow clearly shows that there is still a transfer of the absorbate between the flow over the packing and the wall flow even if the flow is at equilibrium, i.e. if there is no net flow of liquid to the wall or back. This transfer persists as long as there is a concentration difference between the concentration of the absorbate in the wall flow and the flow over the packing in the proximity of the wall (see the first term on the left hand side).

Solution of the above differential equation requires further specification of the initial conditions. This means specification of the concentration in the inlet liquid and gas streams and in the inlet wall flow (unless it is zero). The concentration in the inlet liquid stream need not be in principle constant. It may be specified as a function of the radial coordinate.

4.5. THE EFFECT OF FLOW DISTRIBUTION ON LIQUID HOLD-UP

Nonuniformities of the flow of liquid will necessarily lead to the nonuniformities of distribution of liquid hold-up in the volume of the bed. This is true both for the small and large scale maldistribution of liquid. Liquid flow distribution may thus be also responsible for the differences of hold ups found by various authors.

Experimental studies of Staněk et al. [6] and Staněk and Kolář [11] showed that it is possible to evaluate local values of the hold-up from hold up measurements under various initial distributions of liquid and from the known theoretical distribution of the local densities of irrigation. This approach, of course, reflects the effects of only large scale maldistribution on the total hold up.

Let us consider that local hold up (local in the sense of the diffusion model of flow distribution) within the packing, h_{PL}, is a function of the local irrigation rate, f', that can be expressed alternatively as a polynomial or a power law expression in the following manner

$$h_{PL} = \sum_{i=0}^{k} a_i (f')^i \tag{4-67}$$

$$h_{PL} = h_{STF} + a(f')^p \tag{4-68}$$

The nature of the absolute term h_{STF} in the power law expression will be explained later.

Similarly one can speak of the local hold up of liquid on the wall, h_{WL} (local in the sense of the diffusion model), and depending on the local magnitude of the wall flow, W', again in the form of either a polynomial or a power law expression

$$h_{WL} = \sum_{i=0}^{k'} b_i \left(\frac{W'}{\pi R^2}\right)^i \tag{4-69}$$

$$h_{WL} = b\left(\frac{W'}{\pi R^2}\right) \tag{4-70}$$

We note that these local values of the hold up h_{PL}, h_{WL} cannot be directly measured.

One can, however, measure the hold up in the bed irrigated by a disc distributor and if the radius of the disc and the depth of the bed are not very large the contribution of the liquid present as hold up on the wall will be negligible. In this case the relationship between the hold up of the bed irrigated by the disc, h_D, to the coefficients a_i of the local hold up on the packing will be the following

$$h_D = \sum_{i=0}^{k} a_i I_D(i)\, f_0^i \tag{4-71}$$

$$h_D = h_{STF} + a I_D(p)\, f_0^p \tag{4-72}$$

The moments $I_D(i)$ are defined by means of the known distribution of the dimensionless density of wetting for the disc distributor, f', given in Table 4-1. Thus

$$I_D(i) = \int_0^1 dz \int_0^1 2r\, f_D^i\, dr \tag{4-73}$$

The definition of the non-integer p-th moment is analogous.

The coefficients of the polynomial or the power law expression for the local hold up on the wall can be obtained from measurement when on the top of the bed all liquid is supplied onto the wall. Under such circumstances the wall flow will be very strong, particularly if the depth of the bed is not excessively large. The experimentally obtained hold up in the bed irrigated by the wall distributor, h_W, can be expressed as follows

$$h_W = \sum_{i=0}^{k} a_i\, I_W(i)\, f_0^i = \sum_{i=0}^{k'} b_i\, M_W(i)\, f_0^i \tag{4-74}$$

$$h_W = h_{STF} + a\, I_W(p)\, f_0^p + b\, M_W(s)\, f_0^s \tag{4-75}$$

The definition of the i-th and p-th moments I_W is analogous to that of I_D in Eq. (4-73) except that one uses the distribution of the dimensionless density of wetting, f_W, given by the appropriate solution of the diffusion equation (4-5) for the initial wetting by the wall distributor, see Table 4-1.

The i-th moments of the wall flow are given as

$$M_W(i) = \int_0^1 W_W^i\, dz \tag{4-76}$$

where W_W is the dimensionless wall flow as a function of the dimensionless axial coordinate z given by the solution of the diffusion equation (4-5), see Table 4-1. The noninteger s-th moment of the power law expression is defined analogously.

Staněk and Kolář [11] carried out the underlying experiments necessary for the above outlined analysis for beds packed by 20 mm spheres and 25 mm Raschig rings in a 288 mm in diameter column irrigated by water. The mean densities of irrigation ranged between 0.0003 and 0.013 m s^{-1}.

Experimentally obtained values of h_D and h_W for a series of mean densities of irrigation, f_0, were fitted by polynomials and power law expressions and the appropriate coefficients a, a_i, b, b_i for the local values of the hold up on the packing and the wall were obtained with the aid of the moments I_D, I_W, M_W.

Numerical values of the moments were obtained by numerical integration using densities of irrigation and the wall flow from corresponding analytical solutions.

Data processing using polynomials showed that the absolute term (zero-th moment) is not identical with the static hold up. This, however, came as no surprise as other authors also share the opinion that static hold up, defined as the amount of liquid that remains in the bed after shutting off the liquid feed, has no relevance to the conditions in the irrigated bed.

The absolute term of the polynomials was therefore taken to be *static hold up under the flow regime*, h_{STF}, and used in the power law expression for the local hold up on the packing h_{PL}. The results for the local hold ups correlated by the power law expressions are summarized in Tables 4-2 and 4-3.

The results in these tables can be used to calculate the dependence of the liquid hold-up under various conditions. This was done [11] for the bed irrigated uniformly and a reasonably good agreement was found.

Table 4-2

Coefficients of Eq. (4-68) $h_{PL} = h_{STF} + a(f')^p$.
Local irrigation rate f' is in [mm s^{-1}].

Packing	a	p	Residual root mean square
20 mm spheres	0.927×10^{-2}	0.785	0.224×10^{-3}
25 mm Raschig rings	1.345×10^{2}	0.724	0.283×10^{-3}

Table 4-3

Coefficients of Eq. (4-70) $h_{WL} = c(W'/\pi R^2)^s$.
Local wall flow rate $W'/\pi r^2$ is in [mm s^{-1}].

Packing	b	s	Residual root mean square
20 mm spheres	0.259×10^{-2}	1.069	0.519×10^{-3}
25 mm Raschig rings	0.272×10^{-2}	1.063	0.455×10^{-3}

From the data summarized in the last two tables one can draw the following conclusions: The residual root mean square for the correlation of the local hold up on the wall is approximately twice the value for local hold up on the packing. This is the direct consequence of the fact that with the outlined method the coefficients for the wall hold up are calculated from two experiments (disc and wall distributor), compared to one (only disc) for the hold up on the packing.

The character of the dependences does not differ appreciably for the two packings. The two packings, however, have a very similar size and specific surface area.

The exponent for the local value of the hold up on the wall flow is close to unity and therefore the film flow mechanism (exponent 1/3) does not apply.

For the given conditions an increment of the liquid load will cause a greater increase of the hold up on the packing than on the wall. The liquid on the packing appears to be therefore more "effective" from the standpoint of liquid hold up on the packing than on the wall. This, however, would change particularly with the d_c/d_p ratio.

Nevertheless, computed plots of the total hold up versus mean irrigation rate provide different curves depending on the type of liquid distributor used, on the value of the parameter C (determining the "equilibrium" value of the wall flow) and on the depth of the bed expressed in terms of the parameter H (determining the extent of the departure of the flow distribution within the bed from the conditions given by the distributor toward the "equilibrium" state).

The trends may be characterized as follows: The observed value of the total hold up will be greater (for given f_0) for columns with a lesser tendency to form the wall flow (greater C). Increased depth of the bed (given by H) will yield lower total hold ups if the initial distribution is such that the magnitude of the wall flow increases with the depth (discs, axial nozzles, but also uniform distributors). Increased depth of the bed will yield higher hold ups for given f_0 if the initial distribution is such that the wall flow decreases with the depth of the bed (wall distributor, large diameter annular and ring distributors).

4.6. THE LIQUID FLOW DISTRIBUTION IN THE BLAST FURNACE

The bottom part of the blast furnace, the bosh region, is the place where counter-current flow of liquid and gas occurs. The liquid in this case is molten slag and molten iron descending against the flow of hot blast within a matrix of mostly coke particles.

The flow is particularly complicated as both the liquid and gas are highly nonuniformly distributed. Gas is being fed from the side through tuyeres at very high velocities which normally, considering the liquid rates involved, should lead to the flooding conditions.

Flooding in counter-current flow systems is a very important factor in general as it limits the production capacity of these systems, see e.g. Billet and Schultes [238].

A scheme of the situation is shown in Fig. 4-6 with the gas injection from one side. The liquid, presumably entering the bosh uniformly distributed over the cross section collects at the bottom as a pool periodically emptied by tapping the furnace. The gas is being fed at such high velocities that preclude the presence of liquid in the immediate neighborhood of the tuyere. Accordingly, the "dry region" here appears, the extent of which depends on the flow conditions.

Szekely and Kajiwara [199] investigated the situation both experimentally on a "cold" model and with the aid of a mathematical model. For the latter purposes the whole region of the bosh was divided into the dry zone, considered to be rectangular in contrast to Fig. 4-6 and the two-phase region.

In the dry region the equation govering the distribution of the gas was the differential vectorial Ergun equation written in chapter 3 as Eq. (3-1).

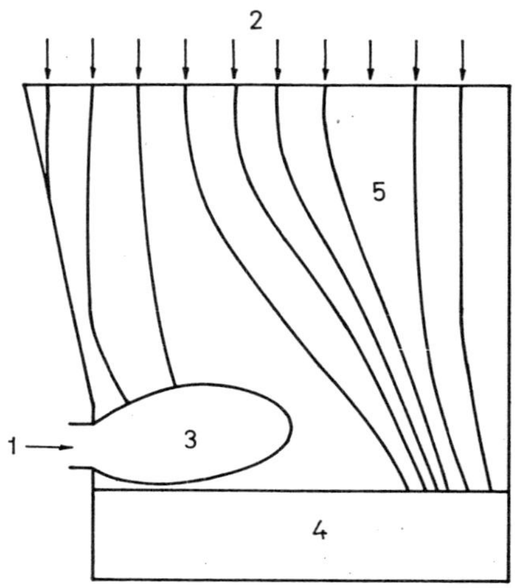

Fig.4-6
Streamlines for the liquid flow in the bosh region of a blast furnace. 1 - gas injection (tuyere); 2 - downward liquid flow; 3- "dry region"; 4 - liquid level; 5 - streamlines.

In the two-phase region, two flow equations were used separately for the flow of gas and liquid. The gas was considered to obey also the differential vectorial Ergun equation, Eq. (3-1) with the coefficients (3-2), (3-3) modified in such a way as to express the reduction of the available cross section (given by the void fraction) due to the presence of the total liquid hold-up, h_t. The coefficients of the vectorial Ergun equation thus take the form

$$f'_1 = 150 \frac{(1 - \varepsilon)^2 \mu_G}{d_p^2(1 - h_t)^3} \tag{4-77}$$

$$f'_2 = 1.75 \frac{(1 - \varepsilon) \varrho_G}{d_p(\varepsilon - h_t)^3} \tag{4-78}$$

The liquid in the two phase region was considered to obey the modified Darcy equation

$$-\text{grad } P_L = f_3 \mathbf{V}_L + \varrho_L \mathbf{g} \tag{4-79}$$

where $\boldsymbol{V}_L$ is the vector of superficial liquid velocity. The acceleration due to gravity, $\boldsymbol{g}$, is also taken as a vector. The resistance coefficient f_3 in this case is defined as

$$f_3 = 180 \frac{(1 - \varepsilon)^2 \mu_L}{d_p^2 h_t^3} \tag{4-80}$$

The boundary conditions used are self evident: the walls are impervious both for gas and liquid. In addition, the velocities at the top of the bosh are taken to be uniform (liquid) or parallel (gas). The profile of the liquid velocity at the bottom was specified on the basis of experiments. The extent of the (rectangular) dry region was also taken over from experiments on a cold model. Solution of the above equations necessitates knowledge of the total (local) hold up as a function of the (local) phase velocities. This was simplified [199] by taking the hold up as a function of liquid rate only

$$h_t = 0.03 + 0.147 V_L^{0.68} \tag{4-81}$$

(V_L is scalar superficial liquid velocity here in cm/s).

We note that under the above statement of the problem the equations for the gas and liquid velocity in the two-phase zone are coupled through the presence of the hold-up but can be solved separately as the dependence of the liquid flow distribution on gas distribution arises only from the variable outlet velocity profile and the extent of the dry region, taken in this case from experimental data.

Based on their calculations Szekely and Kajiwara [199] presented patterns of the stream lines for gas and liquid and the pattern of the contours of constant liquid hold up. Qualitatively the pattern of the gas streamlines is similar to that computed for one phase flow and presented in chapter 3 as Fig.3-4. They found that the gas flow fed from the side drastically affected the pattern of the liquid streamlines that were "pushed" away from the dry zone toward the axis of the furnace. Maximum liquid hold ups were obtained at the limits between the dry region and the "dead man" zone.

Reliable test of the model can best be obtained from experimental observation of the streamlines of the liquid flow. Szekely and Kajiwara showed that measurement of the exit velocities from the bosh region is of little value as it differs little from the uniform flow. Also calculations on the basis of constant liquid hold up, independent of the local conditions, showed relatively weak effect on the pattern of gas streamlines.

Jeschar et al. [195] further showed, using their experimental data, that the conditions at the limits between the dry region and the "dead man" zone correspond to the flooding conditions when one considers correctly the local conditions. The flooding conditions thus appear to control the phase separation process.

On the other hand, when one considers local conditions, these mostly remain under the loading point in the dead man region (considered on the basis of the data for the uniform flow).

List of symbols used to chapter 4

A_n, A_0	coefficients of Bessel function series
B, C	dimensionless parameters of boundary condition Eq. 4-25
c	dimensionless concentration
D	coefficient of radial spread of liquid
d_c, d_p	column and particle diameter
f_G	dimensional superficial velocity of gas
$f = f'/f_0$	dimensionless density of irrigation
f', f_0	dimensional and mean density of irrigation (superficial velocities)
f_D, f_U, f_W	dimensionless density of irrigation under disc, uniform and wall initial distribution
$\mathbf{g}$	acceleration due to gravity
H	dimensionless group defined in Eq. (4-6)
h	depth of bed
h_{PL}, h_{WL}	local hold up on packing and on the wall
h_D, h_W, h_t, h_{STF}	hold up in a bed irrigated by a disc and wall distributor, total hold up, static hold up under flow regime
J_0, J_1, J_n	Bessel function of the first kind, zero-th, first and n-th order
K	wall effect factor defined in Eq. (4-21)
K_p, K_w	permeabilities in the bulk and wall region of bed
$k_L a$	liquid side volume mass transfer coefficient
Mf	maldistribution factor defined in Eq. (4-15)
P_L	liquid pressure
Q	dimensional total flow rate of liquid in the bed
q_n, q_{nj}	roots of characteristic equations
R	column radius
R_F	reflux ratio
$r = r'/R$	dimensionless radial coordinate
r'	dimensional radial coordinate
$\mathbf{V}_L$, V_L, V_r, V_z	superficial liquid velocity vector, its magnitude und components
W'	dimensional wall flow as a volume flow rate

$W = W'/(\pi R^2 f_0)$	dimensionless wall flow
$W(\infty)$	equilibrium wall flow at infinite depth of the bed
z'	dimensional axial coordinate
$z = z'/h$	dimensionless axial coordinate
α	slope of the hypothetical model packing, Eq. (4-43), or relative volatility, Eq. (4-48)
μ_G, μ_L	viscosity of gas and liquid
ϱ_G, ϱ_L	density of gas and liquid
σ	surface tension
φ	angular coordinate
ε	void fraction

Additional symbols defined in text

5. The effect of flow inhomogeneities on heat dispersion in fixed beds

The problem of heat dispersion in a bed of particles passed by a fluid is of considerable importance in connection with a number of industrial plants and processes [32]. An example of such a process is adsorption. Heats of adsorption reach currently very high values and the associated heat transfer affects the underlying process very significantly. From more recent problems one could name the development of hydrogen storage systems on the basis of intermetallides (for review see e.g. [259]), molecular sieve regenerators [307, 308], coal gasifiers [340], and packed bed solar heaters [311]. The theory of heat dispersion is applicable even to such problems as e.g. indirect cooling of computer chips [293].

Heat regenerators with solid beds, common, for example, in metallurgy, utilize the high intensity of gas/solid heat transfer, thanks particularly to the large interfacial surface, and further also the large thermal storage capacity per unit volume, because of the large heat capacity of the solids compared to that of the passing gas.

The shaft of a blast furnace is also essentially a packed bed where heat transfer between gas and very slowly counter-currently moving solids takes place. The problem is complicated in that the size of coke and ore particles is substantially different and calls for use of spatially variable heat transfer and heat dispersion characteristics. An additional complication is the temperature difference between the surface and the core of the particles, this difference being different for the two components of the blast furnace charge (see e.g. Jeschar et al [260]).

From first principles the heat transfer in a bed of particles represents a relatively complicated process with a number of interactions [7, 8] in all phases. Beveridge and Haughey [58] list four mechanisms of heat transfer by conduction, three mechanisms by radiation and three convective mechanisms. Upon taking into account also the effect of the walls containing the bed there exist, according to Gnielinski [273], a total of 14 mechanisms. Some of the above mentioned mechanisms may be frequently neglected, as e.g. transfer by natural convection in gas unless the velocities are extremely low. Similarly, one can neglect also all radiative mechanisms if the absolute temperatures within the bed are low. Even under these simplifying assumptions the problem remains complex because the remaining mechanisms cannot be studied separately. In addition, the coupling of the various mechanisms is both in series and parallel.

For these reasons the heat transfer in a bed of particles passed by a fluid is most often described in terms of the effective conductivity of the bed characterized by the coefficient of effective thermal conductivity. Yagi and Kunii [59] classified the possible mechanisms of heat transfer as those dependent on the velocity of the flow (3 mechanisms) and those independent of the velocity of the flow (4 mechanisms); of the latter, two are convective and two radiative.

The Yagi and Kunii [59] classification has led to a generally accepted approach [60] of considering the effective thermal conductivity as consisting of the contribution of the conductivity of the bed under no-flow conditions and the convective contribution. The convective contribution is associated with the tortuous path of the fluid through the matrix of solids, this contribution being obtainable also from mass dispersion experiments and the analogy between heat and mass transfer as noted already by Baron [339]. A modified correlation for the conductive contribution to the effective radial and axial thermal conductivity was proposed by Specchia and Sicardi [251].

The no-flow contribution is associated mainly with the conductive contribution of the fluid and the particles and the radiative heat transfer. The question of the radiative contribution to heat transfer in packed beds was investigated experimentally e.g. by Kasparek and Vortmeyer [281], Zubrunnen et al [294] and Chiew and Glandt [305]. The literature on the no-flow contribution is very extensive. An exhaustive review of all aspects of this problem has been published by Tsotsas and Martin [291].

Effective thermal conductivity of beds of irregular particles is higher than that of the bed of regular particles [275]. A method of calculating effective thermal conductivity as a sum of convective and conductive contributions for particles of various shapes and particle mixtures was developed by Bauer and Schlünder [282, 283, 302]; for binary mixtures by Okazaki et al [276].

For regular beds there is a possibility of predicting the effective thermal conductivity from a model; for uniformly sized spheres in a hexagonal array this was done by Dietz [274]. Another computational possibility is the Monte Carlo method used e.g. by Sahimi [284].

The problem of heat dispersion in fixed beds was treated in detail in the book by Schlünder and Tsotsas [289].

5.1. THE PSEUDOHOMOGENEOUS MODEL OF HEAT DISPERSION IN FIXED BEDS PASSED BY GAS

Mathematical models of the dispersion of heat in a bed of particles are essentially of two types: One phase (pseudohomogeneous) and two phase models.

Two phase models formulate separately the energy balances in the phases (solid and gas) exchanging heat and therefore appear, at the first glance, more realistic. However, the two phase and the pseudohomogeneous models are experimentally almost undistinguishable if the temperatures of the fluid and the solid phase are close. The magnitude of this temperature difference was analyzed by Wei [295] and conditions were ascertained for the applicability of pseudohomogeneous models. Fixed bed reactors are prone to large hot spots if the heat of reaction is evolved in the solid phase. In moving bed systems [295] maximum temperature differences between phases appear in counter-current flow of equal thermal capacity streams. The same is true for cocurrent flows but the corresponding temperature differences are less.

One phase models view the bed with the flowing fluid as a pseudohomogenous medium and express the balance of energy by a single equation. They are convenient for description of steady state temperature profiles. The pseudohomogeneous models (or sometimes also pseudo-continuum models) exist also in a one-dimensional form (the temperature being only a function of axial coordinate) combining the resistance of the bed (effective radial conduction) and the wall resistance (wall heat transfer coefficient) into a single overall heat transfer coefficient related to the bed average temperature. The relationship between the two-and one-dimensional pseudohomogeneous models and the predictive ability of the former was analyzed by Dixon [292].

The equivalence of the two phase and pseudohomogeneous models, however, is not strictly limited to the case of infinitely large values of gas/solid heat transfer coefficients. In individual cases [61, 62, 280, 338] one can do with a very modest condition of equality of the second derivatives of the temperature of the gas and solid phase in the axial direction also providing that the inlet conditions do not change very rapidly [296].

Heat dispersion models based on the concept of effective radial heat conductivity almost always provide a very good fit of experimental data with theoretical predictions. Changing, however, to a different bed, mass transfer operation or chemical reaction often yields different effective conductivities. The explanation rests in that many of the heat dispersion mechanisms combined into the effective heat conductivity may well be fitted by this concept but their relative contribution and mutual interactions vary from case to case [304]. The ability of the pseudohomogeneous model to describe the packed bed transients was shown by Dixon and Creswell [286] to depend on the choice of the effective parameters.

5.1.1. Heat dispersion characteristics from nonuniform inlet temperature profiles

Dispersion of heat in a cylindrical bed passed by gas at constant mass superficial velocity G_0 and under axial symmetry may be described on the basis of the pseudohomogeneous model by the following equation

$$G_0 c_{\mathrm{pf}} \frac{\partial T}{\partial z} = K_{\mathrm{L}} \frac{\partial^2 T}{\partial z^2} + K_{\mathrm{R}} \left(\frac{\partial^2 T}{\partial r^2} + \frac{1}{r} \frac{\partial T}{\partial r} \right) \tag{5-1}$$

assuming the homogeneous (uniform) profile of the gas velocity. Under conditions prevailing in high pressure liquid chromatography one has to incorporate an additional term expressing heat generation due to deformation of the viscous fluid, i.e. the irreversible conversion of mechanical energy into internal energy. See e.g. Lin and Horvath [244].

The unisotropy of heat dispersion is respected by two different effective heat conductivities of the bed in the "radial" direction, K_R, and in the "axial" direction, K_L.

The model equation (5-1) was solved [25] for a bed of finite depth while the heat transfer across the wall into the surroundings was described by the following boundary condition

$$-K_R \frac{\partial T}{\partial r} = h_w(T - T_0) \quad \text{for} \quad r = R \tag{5-2}$$

where h_w designates the wall heat transfer coefficient into the surroundings of constant temperature T_0.

Conditions at the inlet are specified by the Danckwerts boundary condition which is, in effect, a heat balance about the inlet cross section. If the temperature profile in the gas just before entering the bed, i.e. the forcing function, is given by the function $T_{z=0-}$, generally different from the function describing the temperature profile on the inner side of the inlet cross section, $T_{z=0+}$, their mutual relation is given according to Danckwerts by

$$T_{z=0-} = T_{z=0+} - \frac{K_L}{G_0 c_{pf}} \frac{\partial T}{\partial z} \quad \text{for} \quad z = 0 \tag{5-3}$$

Based on the same argument the gradient of temperature in the outlet cross section in the axial direction is zero. Thus

$$\frac{\partial T}{\partial z} = 0 \quad \text{for} \quad z = h$$

The problem was solved for the case [25, 27, 30, 41] when the inlet forcing temperature profile was a two-value function: Within a concentric circle of radius r_1 the inlet temperature equaled T_1, thus

$$T = T_1 \quad \text{for} \quad 0 \leqq r < r_1 \quad \text{and} \quad z = 0- \tag{5-4}$$

while in the remaining annular part of the cross section the inlet temperature is T_2

$$T = T_2 \quad \text{for} \quad r_1 < r \leqq 1 \quad \text{and} \quad z = 0- \tag{5-5}$$

Details of the solution may be found in [25, 41] and the resulting temperature field within the bed is given by the following expression

$$T^* = \sum_{n=1}^{\infty} \frac{2q_n[(T_1^* - T_2^*)(r_1/R) J_1(q_n r_1/R) + T_2^* J_1(q_n r) J_0(q_n r/R)}{(q_n^2 + Bi^2) J_0^2(q_n)} \cdot$$

$$\cdot \frac{\left(\frac{1}{2} + A_n\right)}{\left(\frac{1}{2} + A_n\right)^2 - \exp\left[-2Pe_{LR}\frac{h}{R}A_n\right] \cdot \left(\frac{1}{2} - A_n\right)^2} \cdot$$

$$\cdot \left\{\exp\left[Pe_{LR}\frac{z}{R}\left(\frac{1}{2} - A_n\right)\right] - \frac{\frac{1}{2} - A_n}{\frac{1}{2} + A_n} \cdot \exp\left[-2Pe_{LR}\frac{h}{R}A_n\right] \cdot \right.$$

$$\left. \exp\left[Pe_{LR}\frac{z}{R}\left(\frac{1}{2} + A_n\right)\right]\right\} \tag{5-6}$$

where A_n designate for simplicity

$$A_n = \sqrt{\frac{1}{4} + \frac{q_n^2}{Pe_{LR} Pe_{RR}}} \tag{5-7}$$

while q_n are the roots of the following characteristic equation

$$q_n J_1(q_n) - Bi\, J_0(q_n) = 0 \tag{5-8}$$

The Biot number, Bi, was defined on the basis of the effective radial heat conductivity as follows

$$Bi = \frac{h_w R}{K_R} \tag{5-9}$$

The Peclet numbers are defined with the radius of the bed as the length scale. Thus

$$Pe_{LR} = \frac{G_0 c_{pf} R}{K_L} \tag{5-10}$$

and

$$Pe_{\mathrm{RR}} = \frac{G_0 c_{\mathrm{pf}} R}{K_{\mathrm{R}}} \tag{5-11}$$

From the standpoint of data evaluation and processing the most important is the outlet temperature profile, i.e. the profile at the level $z = h$

$$T^* = \sum_{n=1}^{\infty} \frac{2q_n[(T_1^* - T_2^*)(r_1/R)\,J_1(q_n r_1/R) + T_2^* J_1(q_n)]\,J_0(q_n r/R)}{(q_n^2 + Bi^2)\,J_0^2(q_n)} \cdot$$

$$\cdot \frac{2A_n \exp\left[Pe_{\mathrm{LR}} \frac{h}{R}\left(\frac{1}{2} - A_n\right)\right]}{\left(\frac{1}{2} + A_n\right)^2 - \exp\left[-2Pe_{\mathrm{LR}} \frac{h}{R} A_n\right] \cdot \left(\frac{1}{2} - A_n\right)^2} \tag{5-12}$$

All temperatures designated by asterisk are dimensionless temperatures related to the temperature of the surroundings, T_0, and expressed as a fraction of the temperature difference $(T_1 - T_0)$. Thus, for instance, for the temperature T we have the following transformation

$$T^* = \frac{T - T_0}{T_1 - T_0} \tag{5-13}$$

In this form the dimensionless tempeature of the surroundings equals zero and the inlet temperature within the radius r_1 is equal to unity.

If the bed is perfectly insulated, i.e. if $Bi = 0$, both infinite series (5-6) and (5-12) must be supplemented by an absolute term $A_0 = (r_1/R)^2\,(T_1^* - T_2^*) + T_2^*$.

By analogous procedure one can derive also temperature profiles existing in the layer for more general inlet temperature profiles. In such cases only a part of the obtained solutions (5-6) and (5-12) would change, namely the part corresponding to the series expansion of the inlet temperature profile.

The solution described of the heat dispersion equation corresponds to a class of heat dispersion measurements based on experimental determination of initially nonuniform temperature profiles of various types. An equally widely spread class of methods is based on measurement of temperature profiles in externally heated beds, see e.g. [68, 272].

5.1.2. The length effect in heat dispersion measurements

The solutions shown above of heat dispersion in beds of particles [27, 30] correspond to the conditions consistent with the aim of evaluating parameters Pe_L, Pe_R and Bi, while the single subscripted Peclet numbers are defined with the particle diameter as the length scale (in the literature sometimes called the Bodenstein number).

In the given experimental arrangement [27, 30] the temperature in the core of the inlet gas stream, T_1, was by 60° to 100° higher than T_2, while T_2 ranged about the usual laboratory temperature, but generally differed from the temperature of the surroundings, T_0.

The parameter evaluation was carried out by optimization of the theoretical outlet profiles, given by Eq. (5-12), and the experimental temperature profiles using the Marquardt algorithm of nonlinear regression [68].

The results of the analysis of sensitivity showed that the method described is useful for the determination of the effective radial thermal conductivities in the form of Pe_{RR} or Pe_R. The radial Peclet numbers can be determined with relatively high precision. The axial thermal conductivities, on the contrary, cannot be evaluated [25, 27]. Under these circumstances, of course, one can use simpler solutions corresponding to the pseudohomogeneous model of a semiinfinite bed with zero axial dispersion.

As a consequence of thermal insulation of the experimental bed the dispersion takes place almost adiabatically. This appears to be advantageous as it partially eliminates effects of structural inhomogeneities of the bed near the walls where the corresponding thermal fluxes are small.

In the same experimental set up with a bed 50 mm in diameter used for the determination of the observability of heat dispersion parameters a large number of experiments were carried out [30] with spherical particles made of steel, lead and glass ranging in diameter from 0.49 to 3 mm. The values obtained of the Peclet number related to the particle diameter, Pe_R, were correlated against the Reynolds number $Re = G_0 d_p/\mu$ and the Prandtl number, Pr, defined on the basis of the specific heat, viscosity and thermal conductivity of the gas, by the following correlation

$$Pe_R = \frac{RePr}{(a + bRePr)} \tag{5-14}$$

Values of the coefficients a, b determined by optimization are summarized for various particles and various depth of the bed in Table 5-1.

The data summarized in this table point to the considerable effect of the length (depth) of the bed , i.e. on something as a "length effect". The intensity of this length effect becomes even more manifest from the graphical representation of the dependence of the radial Peclet number, Pe_R, on the Reynolds number. plotted in Fig. 5-1 for glass spheres 0.78 mm in diameter and various depths of the layer. From this figure one can clearly see the segregation of the data corresponding to different depths. In the given case the length effect is stronger than that of the particle size as may be apparent from the next Fig. 5-2 plotting again Pe_R versus Re for glass spheres 0.49; 0.78; 1.9 and 3 mm measured in all cases on the same depth of the layer equal 85 mm. Virtually no segregation is visible in this case and the data fall practically on a single curve.

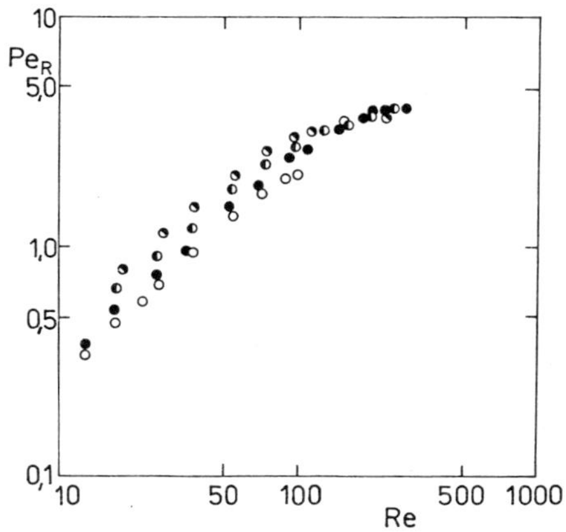

Fig.5-1
Radial Peclet number as a function of Reynolds number for 0.78 mm glass spheres and various depths of the bed. ○ 41 mm; ● 60 mm; ◐ 85 mm; ◑ 112 mm.

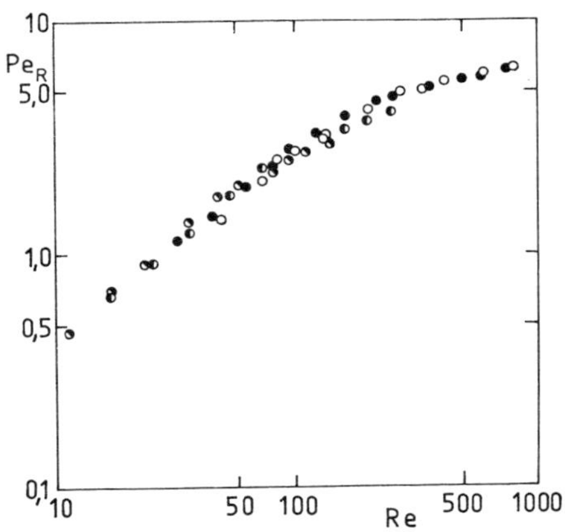

Fig.5-2
Radial Peclet number as a function of Reynolds number for glass spheres of various diameters and 85 mm deep bed. Sphere diameter ○ 3 mm; ● 1.9 mm; ◐ 0.78 mm; ◑ 0.49 mm.

Li and Finlayson [250] detected the length effect manifesting itself in the length dependence of heat transfer characteristics. According to their opinion [250] the length effect appears due to the higher values of heat dispersion characteristics in the inlet region but do not give any detailed explanation for this.

Also Patterson and Carberry [277] observed the length effect and attributed it to the neglect of axial dispersion. Excess fluid flow near the wall was sometimes believed to be the source

Table 5-1

Parameters of the correlation (5-14)

Particles (spheres)	Bed depth [mm]	Parameters **b**	*b*
steel 1 mm	41; 60; 85; 112	49	0.14
steel 3 mm	41; 60; 85; 112	39	0.14
lead 1.75 mm	41	31	0.15
lead 1.75 mm	60	27	0.15
lead 1.75 mm	85	26	0.15
lead 1.75 mm	112	22	0.15
glass 0.78 mm	41	22	0.15
glass 0.78 mm	60	18	0.14
glass 0.78 mm	85	16	0.14
glass 0.78 mm	112	11	0.16
glass 1.9 mm	41	23	0.13
glass 1.9 mm	60	20	0.13
glass 1.9 mm	85	15	0.13
glass 1.9 mm	112	13	0.14
glass 3mm	41	24	0.12
glass 3mm	60	21	0.12
glass 3mm	85	17	0.13
glass 3mm	112	14	0.13

of the length effect by the Taylor dispersion mechanism. Experimental studies, however, do not confirm this hypothesis.

Since axial dispersion proved insignificant also in our experiments [30], it cannot therefore be considered to be the source of the length effect. Since, in addition, the method used for measurement of the change of the inlet temperature profiles requires experiments on relatively shallow beds, eventual effects of axial dispersion should be particularly strong.

On the basis of analysis of results the appearance of the length effect is therefore ascribed to inhomogeneities of the velocity field in the real bed. According to the previous paragraphs these inhomogeneities may appear due to the structural defects or be induced by the inhomogeneity of the temperature field.

Since none of the modifications of the one- or two phase models forsees the dependence of heat dispersion parameters on the depth of the bed, the existence of the length effect to some extent denies the validity of these models.

5.2. THE EFFECT OF STRUCTURAL INHOMOGENEITIES ON HEAT DISPERSION IN FIXED BEDS

Structural effects as a cause of problems in heat and mass transfer in packed beds were identified by Morales et al [263]. For a review see e.g. Tsotsas and Schünder [269] and Ziolkowska and Ziolkowski [143].

A most significant structural inhomogeneity of the bed of equal sized particles is the inhomogeneity near the containing walls of the bed. In chapter 2 we have shown by models as well as experimental results oscillations of the void fraction near the containing surface of substantially larger radius of curvature. The changes of the structure of the bed in this area then induce changes of effective thermal conductivity of the bed for reasons of geometry.

Increased void fraction near the wall further causes preferential flow through this region. This question was investigated in chapter 3 and corresponding velocity profiles may be determined from Eqs (3-27) and (3-28). The changes of the velocity profile bring about additional changes of the effective thermal conductivity of the bed. Flow nonuniformities due to different void fraction near the wall were analyzed theoretically by Carbonell [254] who found that even small differences in the void fraction can lead to significant changes of the dispersion coefficient.

In most cases, where the effective thermal conductivity of the bed is evaluated at various radial positions, it is found to decrease very strongly in the proximity of the wall. This region is thus one of increased resistance to radial heat transfer. A method of incorporating structurally variable effective conductivity of the bed into heat and mass transfer models is via zone models as done by Glicksman and Joos [253], Ziolkowski and Tobis [248] or Martin [88]. Glicksman and Joos found the structural effects to be particularly strong at low Re numbers. The same conclusions have been arrived at by Martin [88] who employed his two zone model with the bulk void fraction for the core of the packed bed and the void fraction given by Eq. (2-48) for the wall region acting as a bypass. Calculations of the particle to fluid heat transfer coefficient (two phase model) using velocities given in Eqs (3-43), (3-44) showed that the influence of structural inhomogeneities on the Nusselt number of the bed as a whole and the apparent bed to fluid heat transfer coefficient for the nonuniform system may become much lower than for a single sphere.

One alternative method of incorporating the increased resistance to radial transfer near the wall is to identify the increased resistance with an apparent bed to wall heat transfer coefficient while the bed as a whole is assigned an effective conductivity typical for the core of the bed. This coefficient is, of course, not identical with real bed to wall coefficients when radial variations of the thermal conductivity are respected. Data that do not reflect the excess heat transfer resistance near the wall are actually bed-averaged values and preclude successful scale up.

The third alternative of incorporating the increased resistance near the wall is to respect radial variations of all quantities: Marivoet et al [133] noted wobbles on radial temperature profiles, observed also by DeWash and Froment [60], and investigated several model aternatives that could possibly explain this behaviour. Mere transition to the oscillatory radial velocity profiles (G_z in Eq. (5-18)) still produced fairly smooth temperature profiles. Wobbles on the temperature profile appeared when the local radial thermal conductivitiy (K_R in Eq. (5-18)) was taken to be a linear function of the local (oscillatory) void fraction, or when both the velocity and the conductivity were taken to have an oscillatory character. Marivoet concluded that the dominating effect causing appearance of the wobbles is that of the radial oscillatory profile of the effective radial thermal conductivity. Local radial effective thermal conductivities and local velocities based on local void fraction measurement were employed by Lerou and Froment [267].

5.2.1. Low bed to particle diameter ratio effects

Structural inhomogeneities become of particular importance in beds of low bed to particle diameter ratio. In fact, at extremely low bed to particle diameter ratios the applicability of the pseudohomogeneous model is rather questionable. This problem becomes one of proper definition of the effective values of the radial (and eventually also axial) thermal conductivity of the bed and the bed to wall heat transfer coefficient.

A method of circumventing these difficulties has so far been to correlate the effective values of the heat transfer characteristics against the d_c/d_p ratio. Rigorous formulation on the basis of the known structural inhomogeneities near the wall (see chapter 2) has not been made available to date.

Dixon et al [245] investigated the effect of extra resistance in the fluid phase for low d_c/d_p ratios while Melanson and Dixon [246] correlated empirically the contribution of the solid phase to the effective conductivity for low d_c/d_p ratios.

Even if the increased resistance near the wall is incorporated by adopting the apparent bed to wall heat transfer coefficient, the radial thermal conductivity still depends on the d_c/d_p ratio. DeWash and Froment [60] proposed the following correlation for cylindrical pellets and the two parameter model (with bed to wall coefficient) in the form with the thermal conductivities in [kcal/m hr deg]

$$K_R = K^0_{\text{noflow}} + \frac{0.0025}{1 + 46(d_c/d_p)^2} Re \tag{5-15}$$

where K^0_{noflow} represents the no-flow contribution.

For the one parameter model the dependence on the d_c/d_p ratio is, of course, stronger and the coefficient of the simplex in the denominator of the last correlation is equal to 120. In both cases, however, the effective radial thermal conductivity increases with the d_c/d_p ratio that is with weakened structural inhomogeneity.

Tsotsas and Schlünder [265] proposed the following expression for the effective radial dispersion of mass in beds of finite bed to particle diameter ratio in terms of the Peclet number, Pe_{finite}, in the form

$$\frac{1}{Pe_{\text{finite}}} = \frac{1}{Pe_{\text{noflow}}} + \frac{1}{Pe_{\infty}(V_0/V_c)} \tag{5-16}$$

where Pe_{noflow} designates the Peclet number based on the effective dispersivity without fluid flow and Pe_{∞} is the Peclet number for d_c/d_p to equal infinity. The superficial velocity of the fluid in the core of the bed, V_c, is lower than the mean velocity V_0 and can be determined from the expressions presented in chapter 2. The last equation offers the possibility of relating dispersion and structural inhomogeneity based on one's own concept of the relationship between the inhomogeneity and velocity.

5.2.2. Structural effects at low and high fluid velocities

At very low Reynolds numbers when the conductive (no-flow) contribution to the effective radial thermal conductivity is dominant the effect of increased void fraction near the confining surface depends on the relative thermal conductivity of the particles and the fluid. If the conductivity of the particles exceeds that of the fluid, the effective thermal conductivity near the wall decreases. If, on the other hand, the conductivity of the fluid is higher, the effective conductivity near the wall increases due to the increased fractional volume occupied by the fluid near the wall. Calculations of this sort based on the void fraction model of Vortmeyer and Schuster [118] were carried out by Ruß [290].

In an effort to eliminate the need for introducing the bed to wall heat transfer coefficient Delmas and Froment [109] formulated three hypotheses for the local variation of the dynamic (flow dependent) part of the effective radial thermal and mass dispersion coefficient dominating at higher fluid velocities. In the first hypothesis, according to the theory of isotropic turbulence, the dynamic contribution is proportional to local superficial velocity. In the second, it is proportional to $(1 - \varepsilon)$ and in the third to the product of the former two quantities. Considerably different and conflicting results were found for the conversion and temperature profiles. The authors warned against combining various hydrodynamic effects into "effective" transport mechanisms particularly for processes with large heat effects.

Dixon and Labua [257] presented the following correlation for the wall to fluid mass or heat transfer coefficient including the effect of the d_c/d_p ratio in the following form

$$Sh = (1 - d_p/d_c)\, Sc^{1/3}\, Re^{0.61} \tag{5-17}$$

The correlation is claimed to be valid fo Re exceeding 50 and particles such as cylinders, hollow cylinders and spheres. The particle diameter, d_p, is calculated as an equivalent diameter of a sphere and hollow cylinders are treated as full.

5.2.3. Wall structural effects – modelling

Dispersion of heat in a layer with structural inhomogeneity near the confing walls is thus governed, from the viewpoint of the pseudohomogeneous model, by the following equation

$$G_z c_{pf} \frac{\partial T}{\partial z} = K_L \frac{\partial^2 T}{\partial z^2} + K_R \left(\frac{\partial^2 T}{\partial r^2} + \frac{1}{r} \frac{\partial T}{\partial r} \right) \tag{5-18}$$

differing from Eq. (5-1) only on the left hand side where the axial velocity component (radially distributed, i.e. $G_z = G_z(r)$) replaced the original (constant) mean superficial velocity, G_0. For this purpose one can utilize expressions from chapter 3, for instance Eq. (3-27), together with $G_z = \varrho V_z$.

Chen et al [258] presented an analytical solution for an axisymmetrical case of unsteady mass dispersion in a packed bed, including radial and axial dispersion term with the axial velocity in the convective term given by the following expression

$$\frac{V_z}{V_0} = a \left[1 - \left(\frac{r}{R} \right)^2 \right] \exp\left[-b \left(\frac{r}{R} \right)^2 \right] \tag{5-19}$$

The axial superficial velocity component, V_z, is thus a function of the radial coordinate. Eq. (5-19) describes, depending on the parameters a, b (b is negative), axially symmetrical flow with off center maxima. The maxima shift towards the axis of the bed as the coefficient b approaches zero.

Solutions obtained by Chen et al [258] indicate that the nonuniform but symmetrical velocity profile strongly influences the local distribution of concentration. The effect of nonuniform axial velocity profile, according to their model, however, becomes small when only the mean concentration, i.e. breakthrough curves are considered. The latter conclusion, of course, is not very encouraging for processes such as chromatography where a corresponding decrease of the number of theoretical plates would be still dramatic. Qualitatively one can say that the lines of constant concentration in the $r - z$ plane, which should be perpendicular to the axis of the bed, change to distorted ellipsoids. The distortion is particularly strong for short elapsed times from injection.

Beasley and Clark [259] carried out a complex study of a packed bed thermal energy storage system. The analysis included both radial and axial dispersion, radial variations of the inlet fluid temperature, spatial variations of the void fraction, wall heat capacity and wall thermal losses. The differential and vectorial form of the Ergun equation (3-1) with variable void fraction was used to establish the velocity gradients governing the radial variations of the axial velocity component. Spatial variations of the void fraction were found to have a very significant effect on the dynamic response of the system both in the fluid and the solid phase.

5.2.4. Wall structural effects in chemical reactors

Models of chemical reactors and experimentally investigated chemical reactions and the role of structural inhomogeneities were reviewed e.g. by Vortmeyer [266]. Choudhary et al [343] investigated the course of a heterogeneous irreversible reaction in a fixed bed exhibiting a concentric cylindrical core with either higher or lower resistance to the flow, see Fig. 3-1. The balance of the reactants took into account both the axial and radial dispersion and the corresponding radial and axial velocity components were obtained from the solution of the differential and vectorial form of the Ergun equation (3-1) for the two dimensional axially symmetric case written in terms of the stream function, Eqs (3-18), (3-19). Since the reaction was assumed to be thermoneutral the system was isothermal throughout and the Ergun and the mass balance equations were uncoupled. The authors [343] solved both equations numerically although, under the above conditions, one can employ analytical solution of the flow equation, Eq. (3-26).

The numerical results obtained can be qualitatively characterized as follows: For the case of the high resistance core, local fluid velocities in the core are lower and higher reaction conversions are achieved closer to the reactor inlet. The discrepancy between position of contours of constant concentration in the core and the annular region of the reactor increases with increasing reaction rate (Damkoehler number). Analogously, if the reactor core exhibits lower resistance to the fluid flow, higher reaction conversions are reached in the core farther away from the inlet in comparison with the annular region of the reactor. However, a bump appears on the contours of constant concentration in the annular region corresponding to lowered resistance due to the wall structural effects superimposed on the generally lower resistance of the annular region. As before, the discrepancy between position of contours of constant concentration grows larger with increasing reaction rate (Damkoehler number). These findings were reasonably confirmed by the experiments of the same authors [344] for the alumina catalyzed isomerization of 1-butene to *cis*- and *trans*-2-butene. The differences of the resistance in the core and the annular region of the reactor 5.25 cm in diameter were created by charging the respective regions with catalyst particles of different size.

Kondelík and Boyarinov [271] found by analyzing a zone model, where the wall zone extended up to the first void fraction minimum, that the conversion of a catalytic reactor as a function of the d_c/d_p ratio displays a minimum for d_c/d_p between 5 and 10 and that the depth of this minimum decreases with increasing radial mixing.

Vortmeyer [261, 297] extended the plug flow dispersion model by taking into account the void fraction and hence the velocity variations caused by structural inhomogeneities of the bed near the wall. The void fraction profiles were computed on the basis of Eq. (2-50) with an exponential decrease. These profiles were substituted into the Ergun equation expanded by the viscous term (3-5) written in one dimensional form with respect to velocity. The resulting expression was solved by variational methods and the effect of the obtained radial profiles of the axial velocity component on processes in fixed beds analyzed.

Vortmeyer concluded that the effects of structural inhomogeneities and hence increased velocities near the wall are more important at low Renolds numbers [341], in wall cooled reactors, [342], and if the axial concentration and/or thermal zones are narrow [261, 297]. In these cases it is difficult for the radial dispersion mechanism to level off the in the wall zone appearing radial concentration and/or temperature profiles. Since, in addition, the dispersion coefficients for mass are substantially lower at low fluid velocities than those for heat the question of structural flow inhomogeneities is of particular concern for reactors and absorbers.

It has been observed that the effective thermal conductivity and the wall heat transfer coefficient in a packed bed change when a chemical reaction occurs. Wijngaarden and Westerterp [312] explained this observation by incorrect application of the pseudohomogeneous model in which the temperatures of the catalyst and the gas are different. This difference causes an apparent dependence of the heat dispersion parameters on the reaction rate and the authors [312] give criteria for which a two phase model must be used.

Sufficient intensity of radial heat dispersion is of particular importance in fixed bed catalytic reactors with highly exothermic reactions. High effective radial thermal conductivity ensures a better stability of the operation and reduces the hazard of runaway conditions. Lee et al [256] achieved increased radial effective conductivity of a fixed bed of cylindrical catalyst pellets by mixing the pellets with 30-39% of steel balls exhibiting substantially higher heat conductivity. Various aspects of runaway conditions in cooled tubular reactors were investigated by Wijngaarden [288].

5.2.5. Other types of structural inhomogeneity effects

Kunii and Suzuki [88] concluded that the decrease of the particle to fluid heat transfer coefficient at low Peclet numbers in beds of fine particles is due to the nonuniform flow (channeling) and therefore uneven contacting of the fluid with the solids. Their model envisages the bed as consisting of aggregates of fine particles with the fluid flowing preferentially through the larger channels between the aggregates.

Cheng and Hsu [114] developed a model predicting the Nusselt number (wall heat transfer coefficient) in an annular sphere packed bed. Structural effects on the void fraction profiles near both cylindrical containing surfaces were described by Eq. (2-50). The Brinkmann equation, linearized by dropping the inertia term, was useful to obtain velocity profiles in an analytical form (see chapter 2). The authors claim a good agreement with the theory based on the mixing length concept and the experimental data of Yagi and Kunii [262]. A vulnerable point of this theory, however, is the inconsistency of using equations on the bed scale as well as on the particle scale.

A similar approach has been followed by Cheng and Vortmeyer [264] to analyze the heat transfer across the packed bed confined between two parallel plates heated from the outside and taking into account the increased heat resistance near the wall.

A theoretical model describing the heat transfer between an immersed heater and the packed bed has been developed by Botteril and Denloye [252] and by Schlünder [298, 301]. The former [252] was based essentially on the two zone model of increased voidage near the heat transfer surface. Predicted results agreed with experimental results and were consistently lower than those based on the constant voidage model.

In the previous chapters we also examined systems consisting of layers of particles of different diameter. Each layer in such a layered bed has thus a different flow resistance. In addition, the dispersion characteristics of such systems are variable depending on the direction of the flow with respect to the orientation of the layers. Corresponding models thus should incorporate directional dispersion, or the tensorial character of the dispersion coefficient. A simplified approach to layered systems, which are viewed as transversely isotropic, was presented e.g. by Moranville et al [249].

5.2.6. Structural inhomogeneities and the length effect

Results of the experiments [30] indicate that structural inhomogeneities of the bed are unlikely to be the cause of the length effect. The extent of the region of increased void fraction, in agreement with chapter 2, is proportional for the given type of particles to d_c/d_p. The results presented in Fig. 5-1 were obtained on the basis of measurements with 0.76 mm glass spheres and hence the d_c/d_p ratio equaled approximately 65. Moreover, the length effect was detected even for 0.49 mm spheres. Also Tobis and Vortmeyer [268] found strong nonuniformity effects in beds in which the above diameter ratio exceeding 100. The hypothesis of the appearance of the length effect due to the structural inhomogeneity of the bed is therefore unlikely.

5.3. THE EFFECT OF THERMALLY INDUCED FLOW INHOMOGENEITIES ON HEAT DISPERSION

The complexity of the dispersion of heat in beds of solids passed by a fluid cannot be elucidated on the basis of structural inhomogeneities only. There is ample experimental evidence pointing to other than structurally induced effects:

Gunn and Khalid [68] found significant differences of thermal characteristics between metallic and non-metallic particles that cannot be accounted for by neglect of axial dispersion.

Roizard and Tondeur [285] showed that the temperature dependence of the densities of the fluid and the solid phase under steady state either spreads or sharpens the temperature wave. These effects may unduely distort the evaluated heat transfer coefficients.

Levec and Carbonell [278] found that effective thermal conductivities under steady state conditions can differ significantly from those observed in transient experiments, due to the spread in temperature pulses. In another paper [279] they attributed to the influence of the density and viscosity gradients the inconsistency of the hydrodynamic contribution effect to dispersion between mass and heat dispersion data.

5.3.1. The effect of thermally induced flow inhomogeneities on heat dispersion in fixed beds

Chapter 3.3 presented the relationships allowing calculation of the velocity field in a layer of particles, which although structurally homogeneous was nonisothermal. These relationships are based on the *P-V-T* behaviour of an ideal gas, Eq. (3-57), whose viscosity, as a function of temperature, however, obeys empirical Eq. (3-58).

Under certain conditions, when the isotherms of the temperature field within the layer of particles are parallel to the direction of the flowing gas, one can find an analytical solution for the axial component of the superficial mass velocity vector, Eq. (3-67), while the radial component, G_r, remains zero. The corresponding pseudohomogeneous model of heat dispersion is then given by Eq. (5-15), i.e. the same equation as that valid for the case of a bed with structural inhomogeneity.

In real sutuations, however, the isotherms for practical reasons cannot be parallel to the velocity vector. Fig. 5-3 depicts the pattern of isotherms in a layer of particles provided that the inlet temperature of the gas is a two-value function of radius, as written in Eqs (5-4) and (5-5), while $T_1 = 353$ K and $T_2 = 293$ K. In such a case the field of isotherms is given by the solution (5-6) evaluated for $Re = 97.5$, $Pe_R = 2.75$. The gas heated to the temperature T_1 enters the bed (see Eq. (5-4)), within the circle of the area equalling 20% of the total bed cross section and hence $r_1 = 0.447R$. The depth of layer of particles in the bed was taken to be twice the column radius. The conditions described thus correspond to the conditions of the experimental determination of heat dispersion and also the value of the radial Peclet number used corresponds to those common for particles studied.

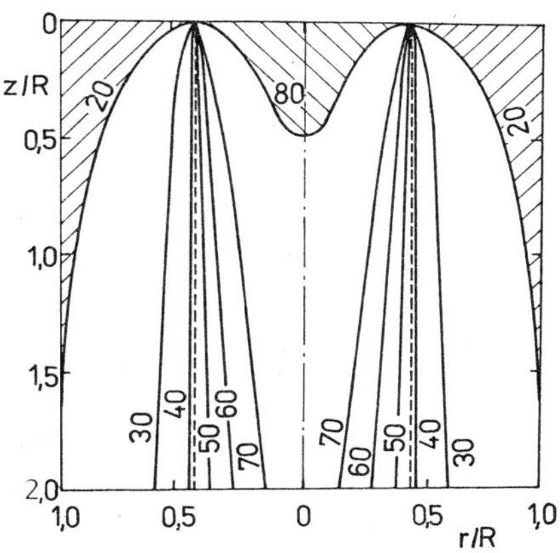

Fig.5-3
Field of isotherms in the *r*-*z* plane of the bed for $T = 80$, $T = 20$, $Re = 97.5$, $Pe = 2.75$, $r = 0.447\ R$, $h = 2R$. Numbers on isotherms indicate temperature in degrees Celsius.

From Fig. 5-3 it is apparent that the isotherms are not parallel to the column axis and hence also not to the presumably prevailing axial direction of the gas flow. The curvature of the isotherms is most conspicuous in the entrance region of the bed, i.e. in the upper part of the figure. The slope of the isotherms changes its sign in the vicinity of the hypothetical cylindrical surface coaxial with the column which divides the cylindrical bed into the part fed by the hot (T_1) and cold (T_2) gas. This hypothetical separating cylindrical surface is shown in Fig. 5-3 by vertical broken lines.

The effect of thermal inhomogeneities on the velocity field, under conditions shown in Fig. 5-3, was investigated by solving Eq. (3-64), while the necessary temperatures were computed from Eq. (5-6) for the above given values of the parameters. Since the effect of axial dispersion was found to be experimentally insignificant the axial Peclet number, Pe_{LR}, was assigned a high value in the calculation.

Solution of Eq. (3-64) was carried out after introducing the stream function in a manner analogous to that leading through the use of Eqs (3-18) and (3-19) to the form (3-21). The boundary conditions for the inlet and outlet cross section of the bed were formulated on the proposition that these cross sections are isobaric surfaces. The streamlines are perpendicular to such surfaces.

The numerical solution proper was carried out by the method of superrelaxation and it produced values for the stream function in individual nodes of the numerical grid. Using these node values of the stream function one can then easily calculate, utilizing the definition expressions, the necessary data regarding the mass superficial velocity vector.

Fig. 5-4 depicts the contours of constant axial component of the gas mass superficial velocity vector, G_z, in the plane passing through the axis of the bed. Corresponding isotherms, we note again, are given in Fig. 5-3. Due to the axial symmetry of the isotherms, these contours are also symmetrical with respect to the bed axis shown by a dash and dot line. The numbers on individual contours express the magnitude of the axial component of mass velocity as a percentage of the mean superficial gas mass velocity, G_0.

From this figure it is apparent that in the vicinity of the bed axis the axial velocity component reaches values lower than the mean. On the other hand, in the peripheral region near the wall the axial velocity component is well above the mean. This observation follows from the fact that owing to the arrangement of the experiment, regions around the bed axis are warmer and hence offer a higher resistance to the flowing gas. The gas therefore redistributes preferentially to regions of lower resistance, that is to colder sections of the bed. Heat dispersion tends to diminish the temperature differences between the axial and the peripheral region of the bed, or, in other words, the temperature profile within the bed becomes progressively flatter. The distribution of the gas mass flow also changes: The axial velocity component in the central section increases with the distance from the bed inlet while in the peripheral section, in contrast, it decreases. Owing

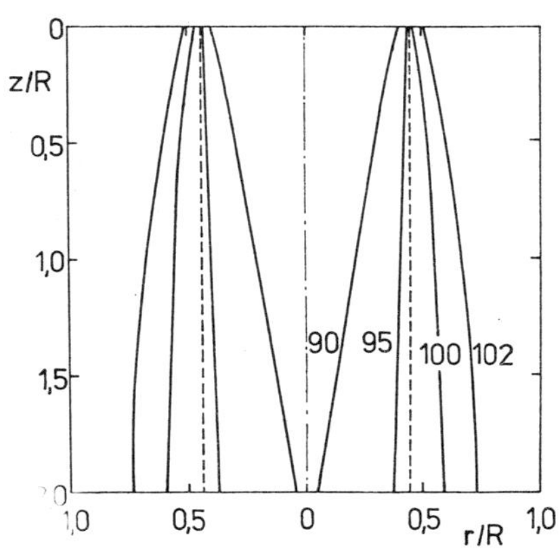

Fig.5-4
Contours of constant velocity component G_z computed from Eq.(3-64) for temperature conditions shown in Fig.5-3. Numbers on contours indicate magnitude of G_z in per cent of G_0 .

to the orientation of the contours of constant G_z as described and the validity of the equation of continuity there must be a nonvanishing radial component of the superficial gas mass velocity vector.

Fig. 5-5 depicts contours of constant radial component of the velocity, G_r, computed from the same values of the stream function as in the previous figure. The depicted component G_r thus corresponds to the temperature field shown in Fig. 5-3. The numbers on individual contours again indicate the magnitude of the component G_r as a percentage of the mean superficial gas mass velocity, G_0.

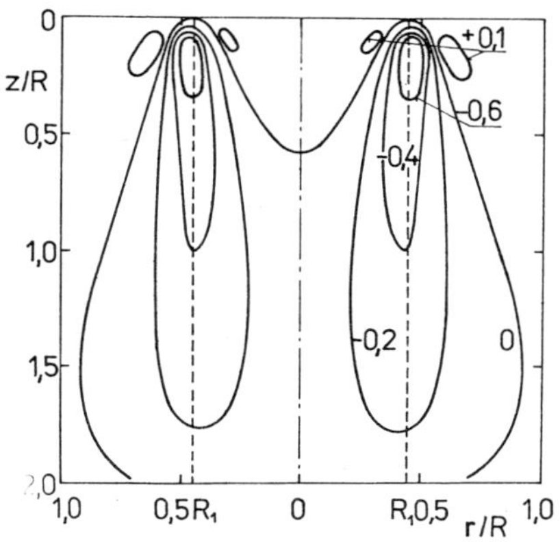

Fig.5-5
Contours of constant velocity component G_r computed from Eq.(3-64) for temperature conditions shown in Fig.5-3. Numbers on individual contours indicate magnitude of G_r in per cent of G_0.

From the figure it is apparent that the radial component of the mass superficial velocity vector assumes mostly negative values characteristic for the flow towards the bed axis. Nevertheless, there exists also a smaller region within the bed exhibiting positive values of G_r and, correspondingly, also the flow toward the wall. Absolute magnitudes of the radial component of the velocity vector are maximal in the region of the hypothetical cylindrical surface separating the region fed by hot and cold gas. This hypothetical surface is shown in Fig. 5-5 by a broken line. Absolute magnitude of the radial velocity component in general also decreases from the maximum in the entrance region in the direction toward the outlet cross section of the bed.

Absolute values of the radial component of the velocity vector are small for the temperature conditions examined and do not exceed 1% of the mean superficial mass velocity of the gas. The relative significance of this component in the overall heat balance may be assessed from comparison of the corresponding convection term with the terms expressing the transport by radial dispersion.

The curve 1 in Fig. 5-6 plots the dependence of the term expressing radial transport in the bed by the dispersion mechanism on the dimensionless axial coordinate. In the dimensionless form this term may be expressed as

$$\frac{K_R}{G_0 c_{pf}} \frac{R}{T_1 - T_0} \left[\frac{\partial^2 T}{\partial r^2} + \frac{1}{r} \frac{\partial T}{\partial r} \right] \tag{5-20}$$

Expectedly, the curve 1 in Fig. 5-6 displays a constantly decreasing trend owing to the diminishing temperature gradients within the bed and hence weakened driving forces for radial heat transfer by dispersion. Corresponding data were again computed from Eq. (5-5) with the above given values of the parameters as was the case in Figs 5-4 and 5-6. The values plotted in Fig. 5-6 were computed for the radial coordinate $r = R/2$.

The curve 2 in the same figure plots the dependence of the radial convective term on the axial coordinate. This term was expressed in the dimensionless form as follows

$$\frac{G_r}{G_0} \frac{R}{T_1 - T_0} \frac{\partial T}{\partial r} \tag{5-21}$$

The radial temperature gradients were calculated as before from Eq. (5-6) for the above given set of parameters while the values of the radial component of the gas mass superficial velocity vector were obtained from Eq. (3-64).

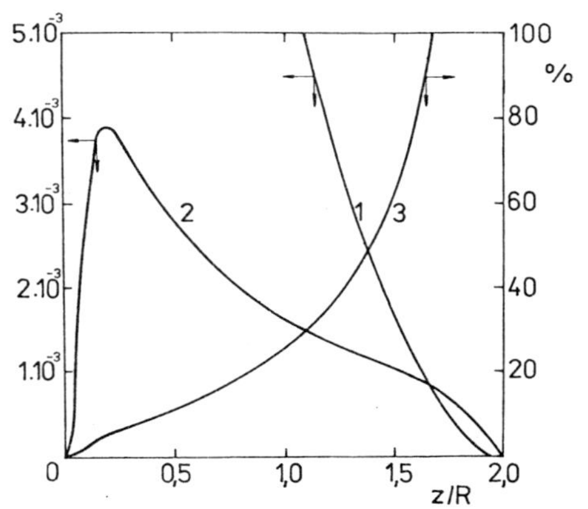

Fig.5-6
Dependence of individual terms of the dimensionless pseudohomogeneous heat dispersion balance on the dimensionless axial coordinate. Curve 1 - radial dispersion term written in (5-20); curve 2 - thermally induced convective term written in (5-21); curve 3 - percentual ratio of the term plotted as curve 2 to the term plotted as curve 1.

As may be seen from Fig. 5-6 the contribution of the radial convective term computed for the radial coordinate $r = R/2$ passes through a maximum in the entrance region of the bed and then steadily decreases. From the course of the curve 3, plotting the ratio of the two previous terms it follows that the importance of the convective term in the overall heat balance steadily increases and for the case analyzed the rate of convective transport exceeds that of the dispersion transfer at the distance of about $z = 1.5R$ from the inlet.

With the experimental technique employed significant deformations of the inlet temperature profile (necessary for a reliable evaluation of Pe_R) occur over relatively short distances of the bed depth. From Fig. 5-6, however, it follows that the contribution of radial convection is virtually never negligible compared to the dispersion mechanism. This is so in spite of the low magnitude of the radial velocity component (see Fig. 5-5).

The equation governing the dispersion of heat thus should take the form

$$G_z c_{pf} \frac{\partial T}{\partial z} + G_r c_{pf} \frac{\partial T}{\partial r} = K_L \frac{\partial^2 T}{\partial z^2} + K_R \left(\frac{\partial^2 T}{\partial r^2} + \frac{1}{r} \frac{\partial T}{\partial r} \right) \tag{5-22}$$

This form of the heat dispersion balance assumes an axially symmetric field of isotherms and hence zero tangential velocity component.

Both components of the velocity vector then must be determined as a solution of the complete Eq. (3-64), or, eventually, some of its asymptotic forms (3-65) or (3-66). In all cases, however, one has to solve these equations simultaneously with Eq. (5-16) as the dispersion of heat depends, through the presence of both velocity components in the heat balance, on the shape of isotherms and in turn the pattern of isotherms depends on the form of the velocity field. The problem at hand is thus clearly a nonlinear one.

Qualitatively one can characterize the role of the thermally induced flow inhomogeneities as follows: The spots within the bed of particles exhibiting a higher temperature offer a greater resistance to the flowing gas owing to the higher viscosity and lower density. As a consequence the gas then flows preferentially through the colder regions.

The importance of the thermally induced flow inhomogeneities can be supported by additional experimental evidence [48], apart from the existence of the length effect in beds with low particle-to-bed diameter ratio: Our experimental results have shown that the length effect grows weaker with increasing thermal conductivity of the particles forming the bed (in the series glass, lead, steel). This can be explained by postulating that the contribution of the mechanisms, that are affected by the gas flow, to heat dispersion decreases with increasing conductivity of the material of particles. This, of course, also diminishes the effect of thermally induced flow inhomogeneities.

From the correlation (5-14) it follows that in the limit $(Re \,.\, Pr) \to \infty$ (i.e. for instance in the limit to very fast flows) the limiting value of Pe_R equals $(1/b)$. On the other hand, in the limit to very slow flows $(G_0 \to 0)$ the effective radial thermal conductivity of the bed is equal to the thermal conductivity of the gas.The values of the parameters a, b, summarized in Table 5-1, are in general lower than corresponding values published in the literature and obtained from experiments with beds heated through the wall. This experimental technique permits measurement with greater bed depths where contributions of the radial velocity component are lower. Accordingly, the values obtained for the effective radial thermal conductivity of the bed are lower. Nevertheless, even these data cannot be taken as asymptotic values either as in reality it is impossible to set-up an experiment with zero axial and nonvanishing radial temperature gradient.

Since the induced convective transport is very intensive compared to the dispersion mechanism, the second term on the left hand side of Eq. (5-22) is significant even in case of relatively small radial components of the velocity vector and hence small nonisothermalities of the bed. This has been demonstrated also by Fig. 5-6 from which the strong coupling of the equation of motion comes. This coupling, without doubt, considerably complicates the modelling of heat transfer in fixed beds passed by gas. Nevertheless, unless all mutual interactions are properly accounted for, data transfer and scale up will necessarily lead to errors.

The role of the coupling between the temperature and velocity fields was examined also by Radestock and Jeschar [76] on the basis of the two phase model of heat dispersion for the case of counter current flow of gas and solids.

Different results of thermal and material transport in isothermal and nonisothermal packed beds were observed also by Schertz and Bischoff [270].

5.3.2. The effect of thermally induced flow inhomogeneities on heat dispersion in fixed bed reactors

In fixed bed chemical reactors the temperature varies not only because of the nonuniformity of the inlet temperature, but, and in fact more frequently, because of the generation of heat by chemical reaction taking place. The cooling of the reactor, removing part of the reaction heat in order to keep the temperature within the prescribed limit,then contributes to the inhomogeneity of the temperature field (nonisothermality). The same situation, of course, applies to endothermic reactions and eventual heating of the reactor in order to keep the temperature and hence the reaction rate at the desired level.

As we have mentioned earlier fixed beds are essentially poor heat conductors in the direction perpendicular to the principal direction of the fluid flow (mostly radial direction). The problem of thermally induced flow inhomogeneities on the dispersion of heat and in turn on the course of the chemical reaction may thus be very important indeed. The effect then becomes manifest through the change of the reactor performance, yield, selectivity and maximum local temperature. All these quantities are important not only from the standpoint of reactor design but also for reactor control and, in some cases, for the safety of its operation.

An attempt to assess the possible extent of thermally induced flow inhomogeneities in chemical reactors has been made by Staněk and Vychodil [241]. The underlying calculations were essentially a numerical solution of the flow equations (3-64) written in terms of the stream function in the form as in Eq. (3-71), i.e. the form accounting for the variations of the density of the flowing gas through the use of the equation of state of an ideal gas, Eq. (3-57), and through the use of an empirical equation (3-58) for the dependence of the viscosity of the gas with its exponent being determined from tabulated values for air within the temperature region of interest.

The governing equation to be solved requires knowledge of the temperature field. For this purpose the authors applied the experimental results obtained at TU München on a laboratory reactor for catalytic oxidation by air of ethane to carbon dioxide. The reactor was a tube 4 cm in diameter packed with catalyst pellets (Pd–γ–Al_2O_3). The inlet reaction mixture contained 0.005 mole/mole of ethane and was preheated to the inlet temperature of 600 K. The temperature of the reactor wall was maintained at the same temperature by intensive cooling and the inlet velocity of gas mixture was 0.1 m s^{-1}. Nominal size of the catalyst pellets was 4 mm and the bed was packed to a depth of 12 cm.

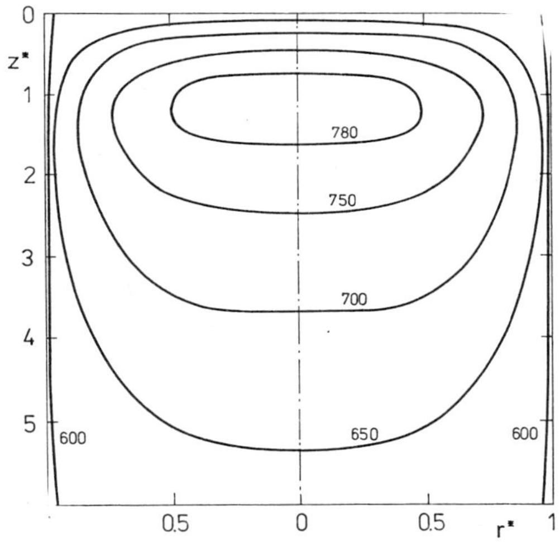

Fig.5-7
Pattern of isotherms in the reactor for oxidation of ethane in the plane passing through the axis. Figures on curves indicate temperature in K. d_c = 4 cm; d_p = 0.4 cm; h = 12; cm; f = 0.1 m/s.

The temperature field obtained is depicted in Fig. 5-7 and reflects the following features. The inlet mixture enters at 600 K and the wall temperature is indeed nearly perfectly 600 K. The temperature of the reaction mixture gradually increases and ultimately exceeds the inlet value by approximately 200 K. The isotherm pattern is axially symmetrical. At a distance approximately one reactor radius the temperature at the axis reaches a maximum and begins to decrease as the reaction rate nears completion. About six radii away from the reactor inlet the temperature across the whole cross section is again close to the inlet value and the temperature of the coolant.

Fig. 5-8 shows the contours of the radial velocity component computed from the streamlines obtained, induced by the temperature field shown in the previous figure. The solution in this case was obtained for the conditions at the inlet dictated by the stipulation that the inlet cross section is an isobaric surface. The figures for individual contours of this figure indicate the radial velocity component as a percentual fraction of the mean gas mass velocity, G_0.

From the figure it is seen that near the reactor inlet the radial velocity component assumes positive values indicating flow toward reactor wall. This tendency is brought about by the fact that the gas tends partly to bypass the hot spot region of maximum reaction temperature through the colder region near the reactor wall. The maximum radial velocity component amounted to about 2% of the mean velocity.

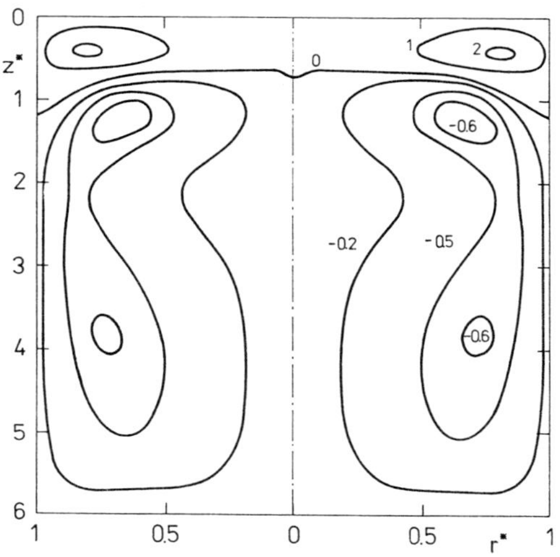

Fig.5-8
Contours of constant radial velocity component computed from the temperature field in Fig.5-7 for isobaric inlet. Figures on curves indicate the magnitude of the radial velocity component in per cent of the mean velocity G_0.

About one radius downstream from the inlet there is the zero radial velocity component contour. Further downstream from this contour the radial velocity component assumes negative values suggesting flow toward the reactor axis. Clearly, the gas tends partially to redistribute back to the central region of the reactor. It is interesting to note that the redistribution toward the axis exhibits two local maxima, both with the radial velocity component about 0.6% of the mean velocity. The same problem was solved also with the inlet conditions assuming uniform feed of the gas mixture over the inlet cross section. The principal difference found was that the maxima of the radial velocity component contours in the inlet region (positive values) were substantially higher and amounted to about 7% of the mean gas velocity.

Fig. 5-9 shows computed [241] contours of the dimensionless axial velocity component obtained again for the isobaric inlet condition. It may be observed that qualitatively the pattern in this figure is similar to that of the isotherms in Fig. 5-7. The minimum axial velocity component is about 87% of the mean G_0 while the maximum exceeds 130%. As expected the minimum axial velocity component is located on the reactor axis while the maximum is in the wall region. Both these observations are in accord with the concept that the thermally induced flow tends partially to avoid hot regions and prefers low temperature passageways.

It is very interesting now, having all this information, to find out what is the importance of these thermally induced flow inhomogeneities relatively in the overall heat balance in the reactor. For this purpose Fig. 5-10 plots the contribution of the radially induced convection (curve 1) and the contribution of the dispersion (curve 2) to net heat transfer. Both these contributions were expressed in the dimensionless form defined similarly as in Eqs (5-20) and (5-21). Curve 3 plots the thermally induced convective contribution as a percentage of the dispersion term. Fig. 5-10 shows these terms as a function of the dimensionless radial coordinate and was evaluated at the axial distance $0.429R$. The plot could not be drawn all the way to the wall due to the numerical difficulties experienced [241] with insufficient accuracy of the second order derivative of temperature from experimental data. The plotted results were obtained for the isobaric inlet condition.

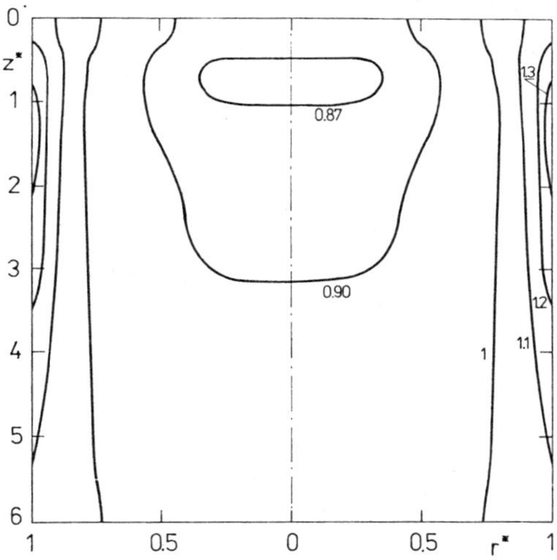

Fig.5-9
Contours of constant axial velocity component computed from the temperature field in Fig.5-7 for isobaric inlet. Figures on curves indicate the magnitude of the axial velocity component as a fraction of the mean velocity G_0 .

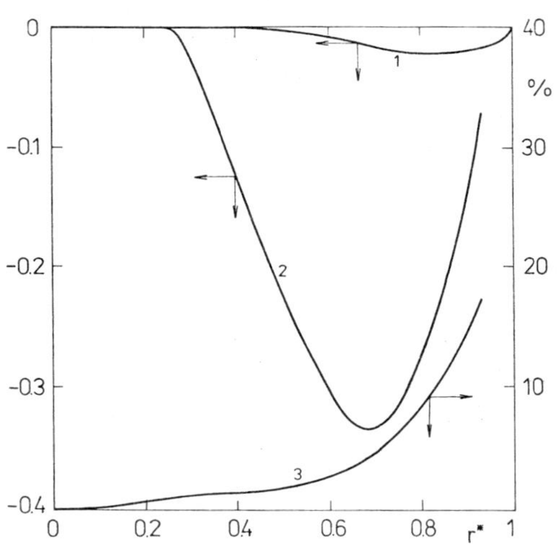

Fig.5-10
The contribution of the thermally induced radial velocity component to radial heat flux at the axial level $z = 0.429R$ computed from the temperature field in Fig.5-7. Curve 1 - thermally induced convective term written in (5-21); curve 2 - radial dispersion term written in (5-20); curve 3 - percentual ratio of the term plotted as curve 1 to the term plotted as curve 2.

The results presented again showed the effect of thermally induced flow inhomogeneities to be strong. It is stressed however that the calculations were only approximate as the iterative procedure applied to the underlying equations involved only iterations of the nonlinear flow equations. In principle, however, the flow equation is coupled with the heat balance and both equations must be solved simultaneously. The results presented, nevertheless, should give a very good estimate as the temperature field used was obtained experimentally.

It is thus seen that for fixed bed reactors the present state of the art does not suffice for a reliable design from the kinetic data and laborious pilot plant experiments are still indispensable. An analysis of what can be achieved by critical modelling of the cooled reactor was presented by e.g. Wijngaarden [288].

It is beyond the scope of this book to analyze the possibilities of the appearance of multiple steady states in fixed bed reactors. Let us note only that apart from the classic case of steady state multiplicities, brought about by nonlinear relationship between the reaction rate and heat transfer rate, there are other possible sources of multiplicities. For certain systems with the same hydraulic resistance several steady states may exist exhibiting different velocities of the flow. These situations are termed thermomechanical multiplicities and may appear due to the high sensitivity of physical properties of the fluid to temperature and structurally affected local changes of the reaction rate. For details the reader is referred to the book edited by Matros [287] and papers by Matros and Chumakova [299] and Boreskov et al [300].

5.3.3. Combined effect of structurally and thermally induced flow inhomogeneities

So far we have examined the effect of structurally and thermally induced flow inhomogeneities separately. In reality, however, both effects would act simultaneously. The principal difference between the two effects is that in case of the structurally induced flow inhomogeneities the flow equation can be solved separately and the resulting components of the velocity vector can in turn be used in the thermal balance, and, in case of a chemical reactor, also in the balance of the reactants.

If the thermally induced flow nonuniformities are also to be taken into account the flow equation has to be solved simultaneously with the heat and mass balances as a set of coupled equations due to the presence of the locally variable temperature in the flow equation. This, of course, results in the necessity for an additional computational effort but there are no principal difficulties in the formulation of the flow equation. In fact, one only has to respect that the resistance coefficients are functions of both the spatially variable structural resistance and the spatially variable temperature.

The governing equation for the three-dimensional case is Eq. (3-50) with the coefficients b_1, b_2 defined, by Eqs (3-48), (3-49), being generally functions of the spatial coordinates through the presence of the viscosity, density, particle size and void fraction. If the flowing gas obeys the equation of state, Eq. (3-57), and its viscosity the power-law expression, Eq. (3-58), the three-dimensional governing equation is Eq. (3-63) with the coefficients c_1, c_2, defined by Eqs (3-60), (3-61), being generally functions of spatial coordinates through the presence of the particle size and void fraction.

For an axially symmetrical two-dimensional case one can solve the underlying flow equation in terms of the stream function defined on the basis of mass superficial velocity components written as Eq. (3-71). The quantity U, appearing in this equation, however, is defined as

$$U = b_1 + b_2 G \tag{5-23}$$

or, if the gas again obeys the equation of state and its viscosity the power law expression

$$U = c_1 \left(\frac{T}{T_0}\right)^{a+1} + c_2 \left(\frac{T}{T_0}\right) G \tag{5-24}$$

where G is the magnitude of the gas mass superficial velociy vector.

Rigorously, in case of chemical reactors one should also account for the change of the physical properties of the flowing fluid due to the changing composition of the reaction mixture. This would further compound to the coupling of the flow equation and the material balance.

In the above paragraphs we have dealt with gas as a flowing fluid where the physical changes come from the temperature variations of both the density and viscosity, the role of the density being stronger in this case. If the flowing fluid is liquid the effect of variable density would be most often negligible but the role of viscosity could increase dramatically, especially in case of high viscosity liquids.

Haidegger and Vortmeyer [225] tested the results of the solutions for the case of the flow equation being coupled and uncoupled from the heat and material balances. The reaction considered was again catalytic ethane oxidation. The material balance was essentially the pseudohomogeneous dispersion with the radial and axial convective term. The heat balance was formulated on the same basis, i.e. with the axial and radial heat dispersion and the heat of reaction term.

The flow equation was the differential vectorial form of the Ergun equation supplemented by the viscous wall friction term, Eq. (3-5), written in the two-dimensional form. Structural flow inhomogeneities were respected by incorporating [225] the radially variable voidage according to the formula of Chandrasekhara and Vortmeyer [83] in Eq. (2-49). In case of "uncoupled" solution this equation was solved for the properties of the gas mixture at constant inlet temperature (600 K). In the "coupled" case the variable properties were varied according to the local conditions.

Haidegger and Vortmeyer [225] found approximately the same additional convective contribution as Staněk and Vychodil [241] stemming from the thermally induced flow inhomogeneities. These additional flow inhomogeneities caused a shift of the field of isotherms downstream with an associated decrease of the temperature maximum by about 7%. Corresponding changes of the field of contours of constant concentration amounted to about 5%.

It is worth noting that in case of exothermic reactions and heat removal through the cooled walls, as in the case just examined, the maximum temperature appears on the reactor axis and hence thermally induced flows prefer paths closer to the wall. Structurally induced flow inhomogeneities also promote stronger peripheral flows. Both effects in the reactor thus compound to a strong net peripheral flow.

On the other hand, an endothermic reaction and the heating of the reactor through the walls will promote stronger axially concentrated thermally induced flows, see e.g. Staněk and Szekely [14], while the structurally induced flows still promote peripheral flows. The two effects thus partly compensate each other to a weaker net flow inhomogeneity.

Comparing the case studies assessing the influence of the thermally induced flow inhomogeneities in the paragraphs 5.3.1 and 5.3.2 for a fixed bed and a fixed bed reactor we note the weaker effect detected in case of the reactor. Elucidation of the roots of this difference helps us also throw some light on the relative importance of the structurally and thermally induced flow inhomogeneities:

Let us note that the calculations for the fixed bed (paragraph 5.3.1) were carried out for the Peclet number $Pe_R = Pe_R(R/d_p)$ which was about 10 times higher than the corresponding value of this criterion for the case of the reactor (paragraph 5.3.2). Pe_R is the Peclet number for the effective radial dispersion of heat (based on the pseudohomogeneous model) with the particle diameter as the length scale. In this definition Pe_R is largely the property of the particles. The differences in the two case studies stem mainly from the effect of the R/d_p ratio. A smaller value of this ratio for the reactor in the case study means more intensive dispersion flux (large particles) over a shorter distance (smaller reactor radius). The net dispersive flux is thus stronger and hence the contribution of the thermally induced convective flux is weaker.

As a rule of thumb one can therefore say that the role of thermally induced radial convective flux would increase with increasing R/d_p ratio, i.e. when smaller particles are used in larger diameter tubes (reactors). This ratio, however, plays a decisive role also in affecting the structurally induced flow inhomogeneities. Structural inhomogeneities, of course, increase with decreasing R/d_p ratio.

Consequently, the structurally and thermally induced flow inhomogeneities have opposite trends in dependence on the aspect ratio R/d_p. In particular, moving to reactors with larger diameter but keeping the same particle size results in the structurally induced inhomogeneities becoming weaker while the thermally induced inhomogeneities increase in importance.

Even for given particles, however, Pe_R, is not perfectly constant and varies also with the Reynolds number. For constant R/d_p ratio the contribution of the structurally induced flow inhomogeneities will change only in the transition region $15 < Re < 150$ (see chapter 3), but within the regions of asymptotically slow $(Re < 15)$ or asymptotically fast $(Re < 150)$ flows will remain constant. Also the contribution of the thermally induced convective fluxes, expressed in terms of the dimensionless velocity components, will change little with Re. More specifically, increasing Reynolds number will cause changes of Pe_R to higher values and therefore will weaken the dispersive flux. This in turn will increase the relative importance of the thermally induced flow inhomogeneities with increasing mass velocity of gas.

List of principal symbols to chapter 5

A_n	coefficients of the series defined in Eq. (5-7)
Bi	Biot number defined in Eq. (5-9)
a, b	parameters of correlation (5-14)
b_1, b_2	resistance coefficients appearing in Eq. (5-23) defined in Eqs (3-48), (3-49)
c_1, c_2	resistance coefficients appearing in Eq. (5-24) defined in Eqs (3-60), (3-61)
c_{pf}	specific heat capacity of fluid
d_c	column diameter
d_p	particle diameter
G_r, G_z	radial and axial superficial mass velocity components
G_0	mean superficial mass velocity
h_w	heat transfer coefficient into surroundings
h	depth of bed
K_L, K_R	effective axial and radial thermal conductivity
Pe_L, Pe_R	axial and radial Peclet number defined on basis of d_p
Pe_{LR}, Pe_{RR}	axial and radial Peclet number defined on basis of R
Pr	Prandtl number defined on basis of properties of fluid
q_n	roots of Eq. (5-8)
R	bed radius
Re	Reynolds number defined on basis of d_p and G_0
r	dimensional radial coordinate

r_1	dimensional radius
Sc	Schmidt number
Sh	Sherwood number
T, T^*	dimensional and dimensionless temperature defined in Eq. (5-13)
T_1, T_2, T_1^*, T_2^*	dimensional temperatures within and outside radius r_1 and their dimensionless counterparts
V_0, V_z, V_c	mean superficial velocity, its axial component and its value within bed core
z	dimensional axial coordinate
ε	void fraction

Additional symbols are defined in text.

6. The effect of flow inhomogeneities of trickling liquid on the course of a heterogeneous catalytic reaction

Trickle bed catalytic reactors are to provide mutual effective contact of liquid and eventually also gaseous reactants with the active surface of the solid catalyst. Early information about catalytic reactions in trickle bed reactors may be found in patents [63, 64] from the beginning of the 1950's. At present their main industrial application is the hydrotreatment of various petroleum fractions: hydrodesulphurization of heavy oils, liquid phase hydrocracking, hydrotreatment of lubricating oils. Other examples include synthesis of butynediol, hydrogenation of alpha-methylstyrene, treatment of waste waters. Reaction systems are usually strongly exothermic (hydrogenations, oxidations). Numerous reviews concerning the trickle bed reactors have been published in the past [201, 329, 330, 331, 334].

A trickle bed reactor represents essentially a special case of a packed bed system, mostly with the cocurrent flow of the liquid and the gas phase. Difficulties with its exact description as well as the comprehension of certain observed phenomena are associated with the multiplicity of the mechanisms of heat and mass transfer and their mutual interactions, including those with the fluid flow and the chemical reaction taking place.

The behaviour of trickle bed systems is therefore very complex indeed. This, no doubt, is a serious obstacle to the design of trickle bed reactors on the basis of mathematical models. The possible multiplicity of hydrodynamic states further compounds the complexity of trickle bed systems. Lazaroni [332] attributed the multiplicity of hydrodynamic states to partial wetting of the particle surface. Under such circumstances a large number of particles are wetted when liquid flow is increased thus generating "new" liquid paths that survive when the liquid rate is reduced to the initial value.

This hysteresis is probably associated also with the different dynamics of the hydrodynamics; Jiřičný and Staněk [333] demonstrated experimentally that the rate of transition from one hydrodynamic state to another is different depending on whether the velocity change (liquid or gas) that caused the change of the hydrodynamic state was higher or lower than that corresponding to the original state. However, multiplicity and the problem of dynamics of trickle bed systems are beyond the scope of this book.

Gutsche et al [328] investigated the hydrodynamics (hold up), heat transfer (radial heat dispersion and wall heat transfer coefficient) and axial dispersion coefficient in trickle bed reactors and the ability of the methods valid for one phase flow to predict the condition in trickle beds. Well developed theory and techniques exist for RTD measurement but Shapiro and Brenner [325] expressed doubts about the utility of the residence time distribution theory for design of chemical reactors at least in circumstances where the reactivity is locally inhomogeneous.

In extreme situations mutual interactions of the velocity fields, reaction rates and thermal fluxes lead to the appearance of undesirable reactions, disproportionations, appearance of dry catalyst spots and eventually to complete evaporation of the reaction mixture with characteristic hot spots.

To the common quantities through which one can control the reactor, such as e.g. inlet concentration, inlet temperature of the feed, intensity of cooling and the quantities that may be adjusted through reactor design, such as e.g. size and shape of particles, reactor diameter, one can thus also add the quantities affecting the homogeneity of the velocity field of the reactants.

The inhomogeneities of the gas flow and their relationship to the structure of the bed were treated in chapter 3. Since, however, the gaseous reactant is usually in surplus or virtually pure (hydrogen), the more important effect in the most widely used trickle bed reactors is that of the homogeneity of the liquid flow.

For gas-limited reaction Baudry et al [323] found the liquid downflow to give usually better conversion than the liquid upflow in the bubble flow regime.

6.1. THE EFFECT OF FLOW INHOMOGENEITIES OF TRICKLING LIQUID ON THE BEHAVIOUR OF AN ADIABATIC TRICKLE BED REACTOR

Trickle bed reactors are generally poor heat conductors. Large scale reactors may be therefore considered as adiabatic. Nevertheless, radial heat transfer characteristics are important because near the wall the heat losses are not negligible and the liquid flow maldistribution is responsible for the heat conductivity maldistribution. Crine [71] analyzed the interactions between bed hydrodynamics and heat transfer properties both at the particle and the bed scale level using the percolation theory concept outlined in chapter 4.

Farid and Gunn [316] measured radial and axial dispersion in trickle beds under conditions such that the dispersion coefficients are not affected by the wall flow. These values were generally lower than those found by other investigators from the estimates of overall dispersion including the effect of the wall flow.

They inferred that the dispersion can be expressed as local radial and axial dispersion within the bulk bounded by the conditions for the flow and concentration at the wall.

As has been described in chapter 4, the trickling flow of liquid through a layer of particles deviates from the uniform flow as the liquid is not always applied evenly to the top of the column and, even if it is, the original uniform distribution gradually changes due to the formation of the wall flow.

Colombo et al [315] showed that internal wetting (pore filling) of the catalyst in a trickle bed reactor is practically complete even at very low liquid rates, probably due to the capillarity. This seems to justify the use of pseudohomogeneous model approach and the same is true for industrial reactors provided that there is a good macroscopic liquid distribution and the reaction is not highly exothermic. Either of these conditions can lead to the appearance of large unirrigated zones in the former case due to the flow maldistribution, in the latter due to liquid evaporation. Evaporation in terms of the diffusion model approach will be treated later in this chapter.

In the case of a trickle bed reactor the liquid carries substrate undergoing a usually exotermic chemical reaction on contact with the catalyst surface. In this chapter we shall again use the diffusional pseudohomogeneous model to describe the inhomogeneities of the flow of liquid and the wall flow on the course of the chemical reaction. The situation is shown schematically in Fig. 6-1.

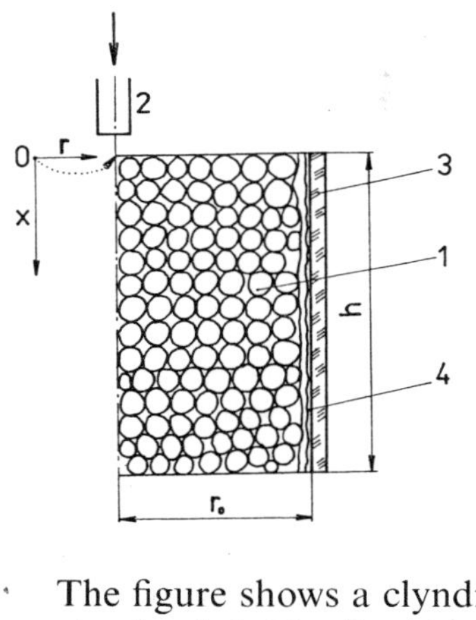

Fig.6-1
Sketch of a trickle bed reactor and associated system of coordinates. 1 - catalyst bed; 2 - feed nozzle; 3 - reactor wall; 4 - wall flow.

The figure shows a clyndrical reactor of radius r_0, filled with particles (1) up to the depth h. The liquid is brought into the reactor by a distributor (2), shown as a central jet of small radius. By the mechanism of random dispersion the liquid during its descent spreads also radially and reaches ultimately the wall (3) where it forms the wall flow (4) with axially variable magnitude.

Owing to the nonuniform initial irrigation of the bed and the formation of the wall flow the local density of irrigation of the catalyst is generally a function of spatial coordinates. The dimensionless density of irrigation, f, expressed as a fraction of the mean density of irrigation, f_0, is given, for an axially symmetrical case by the solution of Eq. (4-5).

In this chapter we shall use also a dimensionless system of coordinates, shown in Fig. 6-1, in which the coordinate of radius, r, is expressed as a fraction of the reactor radius, r_0, while the axial coordinate, x, is defined as a fraction of the depth of the catalyst bed, h. For the dimensionless density of irrigation we may therefore use Eq. (4-5) while the dimensionless quantity, H, is defined again as

$$H = \frac{Dh}{r_0^2} \tag{6-1}$$

The quantity H may regarded as another form of the dimensionless depth of the bed.

The condition of axial symmetry

$$\frac{\partial f}{\partial r} = 0 \quad \text{for} \quad r = 0 \tag{6-2}$$

follows from the assumption of axially symmetric feed (distributor) on top of the layer.

Eq. (4-5) represents a balance on all liquid brought onto the top of the catalyst bed. There is separate balance for the liquid flowing down the reactor wall, i.e. the wall flow. The magnitude of the wall flow in the dimensionless form, W, is expressed as a fraction of the total flow rate through the reactor, $\pi r_0^2 f_0$, and may be expressed from the macroscopic balance by

$$W = 1 - 2\int_0^1 rf \,\mathrm{d}r \tag{6-3}$$

expressing the fact that the sum of the wall flow and the total flow over the surface of the catalyst is constant on any horizontal cross section.

An alternative way of expressing the wall flow utilizes the balance on the liquid exchanged between the wall flow and the flow over the catalyst. The flux of liquid from the catalyst bed into the wall flow is expressed in the dimensionless form by the boundary condition (4-25). The increase of the wall flow expresses mathematically the following equation

$$-\frac{\partial f}{\partial r} = B(f - CW) = \frac{1}{2H}\frac{\mathrm{d}W}{\mathrm{d}x} \quad \text{for} \quad r = 1 \tag{6-4}$$

The net balance on liquid, given by the parabolic differential equation (4-5), may be solved independently of other balances and, in addition, in the analytical form. A general form for an arbitrary axially symmetrical distributor on the top was already written in Eqs (4-28)–(4-30).

The liquid trickling down the catalyst surface in the reactor contains substrate A, which on contact with the catalyst undergoes an irreversible chemical reaction according to the following scheme

$$\mathrm{A} \rightarrow \mathrm{B}$$

This reaction is generally of the n-th order with respect to the substrate.

For the formulation of the balance on the substrate A we shall postulate the following assumptions:

1) The gas phase, stagnant or flowing, participating or not in the reaction, does not affect the transport phenomena within the catalyst bed.

2) Axial dispersion of the reactant is negligible. (Goto and Smith [314] found from experiments in a trickle bed reactor for oxidation of formic acid at 40 atmospheres axial dispersion to be less important than gas-to-liquid mass transfer and intraparticle diffusion)

3) The mechanism of radial dispersion of the reactant is the same as that of radial dispersion of liquid as a whole; other types of transport of the reactant as e.g. molecular diffusion are negligible.

4) The mechanism of exchange of the reactant between the liquid trickling down the catalyst surface and the wall flow is the same as that of the exchange of liquid as a whole.

5) The surface of the wall has no catalytic effect. With these mostly well acceptable assumptions one can formulate the balance on the reactant A in terms of its dimensionless concentration, c, expressed as a fraction of the inlet concentra tion, c_0, in the form

$$\frac{\partial(fc)}{\partial x} - H\left[\frac{\partial^2(fc)}{\partial r^2} + \frac{1}{r}\frac{\partial(fc)}{\partial r}\right] + \frac{Rh}{c_0 f_0} = 0 \tag{6-5}$$

Symbol R designates the rate of conversion of substrate in unit volume of the reactor, which is, of course, a function of temperature and concentration. The first term on the right hand side of Eq. (6-5) expresses axial transport of the reactant A by convection with liquid. The second and the third term express radial dispersion of the reactant A carried by the liquid. Its concentration, c, therefore will be a function of both coordina tes r, x.

Thee bounday condition for Eq. (6-5)

$$-\frac{\partial(fc)}{\partial r} = B(fc - CWc_{\mathrm{w}}) \quad \text{for } r = 1 \tag{6-6}$$

expresses (on the left hand side) in the dimensionless form the flux of the reactant A into the wall flow as (on the right hand side) a product of the driving force for the process and the coefficient B. The driving force is the difference of the product of the density of irrigation and concentration near the reactor wall and the product of the wall flow with the concentration of reactant A in the wall flow, c_w, and the coefficient C. In view of the assumption 4) the parameters of this boundary condition are identical with those of Eq. (4-25).

We note that with this formulation of the boundary condition there is an exchange of the reactant A between the wall flow and the flow over the catalyst surface even in the case where both flows are at dynamic equilibrium. Under such circumstances (see the Table 4-1) we have that $f = C/(1 + C)$ and $W = 1/(1 + C)$ and eventual concentration difference between $c(r = 1)$ and c_w on the given height x is equalized by the mechanism of the dynamic exchange of liquid between the wall flow and the flow down the catalyst surface.

The condition of axial symmetry, requiring zero derivative of the quantity (fc) at the reactor axis reduces, in view of Eq. (6-2), to

$$\frac{\partial c}{\partial r} = 0 \quad \text{for} \quad r = 0 \tag{6-7}$$

The need for an equation for the concentration c_w is met by formulating the balance on the reactant A in the wall flow. This can be easily formulated with the aid of the boundary condition (6-6) and the assumption 5) that the reactor wall has no catalytic effect. Thus we obtain

$$\frac{\mathrm{d}(Wc_w)}{\mathrm{d}x} = -2H\frac{\partial(fc)}{\partial r} = 2HB(fc - CWc_w) \quad \text{for} \quad r = 1 \tag{6-8}$$

Since heterogeneous catalytic reactions are usually not thermoneutral and the reaction term in Eq. (6-5) depends strongly on temperature, one has still to formulate the enthalpy balances enabling the temperatures T and T_w to be computed.

The enthalpy balances are formulated on the following assumptions:

1) The gas phase, stagnant or flowing, does not contribute significantly either to the radial or axial transport of heat in the bed.

2) Axial dispersion of heat is negligible.

3) The mechanism of radial dispersion of heat is the same as that of radial dispersion of liquid as a whole.

4) The mechanism of exchange of heat between the liquid trickling down the catalyst surface and the wall flow is the same as that of the exchange liquid as a whole.

5) The surface of the wall has no catalytic effect.

6) The heat capacity of the gas is negligible.

The above list of assumptions is not complete; some self evident ones, such as constancy of physico-chemical properties, follow from the following equations.

Assumptions 1) -5) are analogous for the heat and mass balances. The key assumptions are 3) and 4), identifying the radial transport of the heat and reactant with that of liquid as a whole. Already at this stage it can be said that while this assumption in case of the mass balance is reasonable, its validity for heat dispersion is only an approximation. In fact, radial thermal conductivity in trickle bed reactors is made up of two components, one due to the heat conduction and the other due to radial liquid dispersion. Chu and Ng [321] evaluated the former on the effective medium theory and the latter on the random dispersion analysis.

Nevertheless, for the time being we shall assume that only radial liquid dispersion is the controlling mechanism of heat transfer. This assumption will be relaxed in paragraph 6.2.

The dimensionless form of the enthalpy balance in the liquid trickling down the catalyst surface takes then the following form

$$\frac{\partial(fT)}{\partial x} - H\left[\frac{\partial^2(fT)}{\partial r^2} + \frac{1}{r}\frac{\partial(fT)}{\partial r}\right] - (-\Delta H)\,\gamma\frac{Rh}{f_0\, c_p\, \varrho_0} = 0 \tag{6-9}$$

The dimensionless temperature, T, was defined using the activation energy, γ, and the Frank-Kamenetskii transformation

$$T = \frac{\gamma}{T_0}(T' - T_0) = \frac{E}{R_g T_0^2}(T' - T_0) \tag{6-10}$$

In the same way we shall define also the dimensionless temperature of the wall flow, T_w. The balance of heat in the wall flow, starting from the assumption 4) and the assumption that the wall is thermally insulated so that there are no heat losses there, may be written as follows

$$\frac{\mathrm{d}(WT_w)}{\mathrm{d}x} = -2H\frac{\partial(fT)}{\partial r} = 2HB(fT - CWT_w) \quad \text{for } r = 1 \tag{6-11}$$

Catalytic reactions taking place will be considered to be generally of the n-th order with respect to the reactant A. Under isobaric conditions and under conditions that the external resistance to mass transfer is negligible the reaction term, R, in Eqs (6-5) and (6-9) may be expressed by the following formal kinetic equation

$$R = k(c\,.\,c_0)^n \tag{6-12}$$

where the kinetic constant depends on temperature by the usual relationship

$$k = k' \exp\left[\frac{E}{R_g T'}\right] = k_0 \exp\left[\frac{\mathrm{T}}{1 + \mathrm{T}/\gamma}\right] \tag{6-13}$$

where k_0 designates the reaction rate constant at temperature T_0.

The above formulated model may be classified as a radially distributed pseudohomogeneous model.

In the balance equations (6-5), (6-8), (6-9) and (6-11) appear products of the dependent variables f, W, c, c_w, T, T_w, while f and W may be obtained from independent solutions given in chapter 4. Before proceeding further it is therefore advantageous to separate the dependent variables utilizing the relationships valid for f and W.

In case of the mass balance on the reactant A we then obtain

$$f\frac{\partial c}{\partial x} - H\left[f\frac{\partial^2 c}{\partial r^2} + 2\frac{\partial f}{\partial r}\frac{\partial c}{\partial r} + \frac{1}{r}f\frac{\partial c}{\partial r}\right] + Da_n c^n \exp\left[\frac{T}{1 + T/\gamma}\right] = 0 \tag{6-14}$$

together with the initial condition

$$c = 1 \quad \text{for} \quad x = 0 \quad \text{and for} \quad 0 < r < 1 \tag{6-15}$$

and the boundary condition (6-7) and

$$-f\frac{\partial c}{\partial r} - c\frac{\partial f}{\partial r} = B(fc - CWc_w) \quad \text{for} \quad r = 1 \tag{6-16}$$

The initial condition (6-14) implies that the substrate is fed at concentration c_0.

The symbol Da_n designates the Damkoehler number, defined for an n-th order reaction

$$Da_n = \frac{R_0 h}{c_0 f_0} = \frac{k_0 c^{n-1} h}{k_0} \tag{6-17}$$

where R_0 designates the reaction rate at concentration c_0 and temperature T_0.

The separation of temperature and the density of irrigation leads to

$$f\frac{\partial t}{\partial x} - H\left[f\frac{\partial^2 T}{\partial r^2} + 2\frac{\partial f}{\partial r}\frac{\partial T}{\partial r} + \frac{1}{r}f\frac{\partial T}{\partial r}\right] - Da_n c^n U \gamma \exp\left[\frac{T}{1 + T/\gamma}\right] = 0 \tag{6-18}$$

with the initial condition

$$T = 0 \quad \text{for} \quad x = 0 \quad \text{and for} \quad 0 < r < 1 \tag{6-19}$$

signifying that all liquid is fed into the reactor at temperature T_0. The boundary conditions read

$$\frac{\partial T}{\partial r} = 0 \quad \text{for} \quad r = 0 \tag{6-20}$$

$$-f\frac{\partial T}{\partial r} - T\frac{\partial f}{\partial r} = B(fT - CWT_\mathrm{w}) \tag{6-21}$$

The balance on the reactant and the enthalpy in the wall flow assume after separation the following form

$$2HB(fc - CWc_\mathrm{w}) - c_\mathrm{w}\frac{\mathrm{d}W}{\mathrm{d}x} - W\frac{\mathrm{d}c_\mathrm{w}}{\mathrm{d}x} = 0 \tag{6-22}$$

and

$$2HB(fT - CWT_\mathrm{w}) - T_\mathrm{w}\frac{\mathrm{d}W}{\mathrm{d}x} - W\frac{\mathrm{d}T_\mathrm{w}}{\mathrm{d}x} = 0 \tag{6-23}$$

with the initial conditions

$$c_\mathrm{w} = 1 \quad \text{and} \quad T_\mathrm{w} = 0 \quad \text{for} \quad x = 0 \tag{6-24}$$

signifying that the wall flow on the top has a concentration of A c_0 and temperature T_0.

The quantity U, appearing in Eq. (6-18), is a dimensionless adiabatic temperature rise defined as

$$U = \frac{(-\Delta H)\, c_0}{T_0\, \varrho\, c_\mathrm{p}} \tag{6-25}$$

Eqs (6-14), (6-18), (6-22) and (6-23), together with the solution (4-29) of Eq. (4-5) and the initial and boundary conditions represent a full statement of the problem of the course of the heterogeneous catalytic reaction in a trickle bed reactor under general inhomogeneities of irrigation within the bed.

In the limiting case when the parameter C of the boundary conditions assumes infinite vales and if the irrigation of the top of the bed is uniform, solution of Eq. (4-5) is trivial: $f = 1$ and $W = 0$ for all x and r. The problem thus reduces to one of the plug flow when the temperature and concentration are constrained by the well-known invariant. In the nomenclature of this work this invariant takes the following form

$$T + U\gamma c = U\gamma \tag{6-26}$$

Upon factoring Eq. (6-9) by the product $U\gamma$ and adding the result with Eq. (6-5) the following partial differential equation results

$$\frac{\partial P_p}{\partial x} - H\left[\frac{\partial^2 P_p}{\partial r^2} + \frac{1}{r}\frac{\partial P_p}{\partial r}\right] = 0 \tag{6-27}$$

in which for brevity the quantity P_p designates

$$P_p = f\left(c + \frac{T}{U\gamma}\right) \tag{6-28}$$

Similarly if we divide Eq. (6-11) by the product $U\gamma$ and add the result to Eq. (6-8) we obtain the following ordinary differen tial equation

$$\frac{\mathrm{d}P_w}{\mathrm{d}x} = -2H\frac{\partial P_p}{\partial r} = 2HB(P_p - CP_w) \tag{6-29}$$

where the quantity P_w is defined analogously to Eq. (6-28)

$$P_w = W\left(C + \frac{T_w}{U\gamma}\right) \tag{6-30}$$

Ii is apparent that the differential equations (6-27) and (6-29) are formally identical with Eqs (4-5) and (6-4) for the quantities f and W. If we further note that also with the initial conditions $T = T_w = 0$ and $c = c_w = 1$ also the initial conditions for the quantities P_p, P_w are idential with those for f and W, we can write the following identity

$$P_p \equiv f \text{ and } P_w \equiv W \tag{6-31}$$

Combination of this result with the definitions (6-28), (6-30) yields that the invariant (6-26) holds also for the local quantities c, T obeying the pseudohomogeneous radially distributed model. Equally valid is an analogous invariant for the local quantities in the wall flow

$$T_w + U\gamma c_w = U\gamma \tag{6-32}$$

From the practical standpoint one can use this solution in such a way that the set of differential equations shown above is solved for only one of the pair of the dependent quantities, concentration/temperature, while the remaining one is computed from the invariants (6-26), (6-32).

From the invariant (6-26) it further follows that at complete conversion of the substrate ($c = 0$) the local temperature in the adiabatic reactor will reach the dimensionless temperature $U\gamma$. This product thus represents the maximum attainable temperature in the reactor under adiabatic conditions.

Since the balance on the total amount of liquid and the wall flow are not coupled with the solution of the balance equations on the reactants, its solution may always be obtained in advance in the form of a series of the Bessel functions (see chapter 4). The remaining two partial and two ordinary differen tial equations, however, are coupled. More details about the method of their solution may be found in [24].

Let us note that the existence of the invariants (6-26) and (6-32) was not made use of, as suggested above, for the economy of the solution. It turned out useful to solve both the material and the energy balances and use the two invariants for checking the internal consistency of the numerical results.

Numerical solutions were obtained for the following specifications:

Liquid distributors: Uniform distributor irrigating the whole top cross section of the catalyst bed. Disc distributor irrigating uniformly the top cross section within the radius r_1 with r_1 being equal 0.1 and 0.5 of reactor radius. Annular distributor irrigating uniformly the top cross section within a ring between radii r_1 and r_2 while taking these radii as 0.5 and 0.6 multiples of the reactor radius respectively.

Parameters of liquid distribution: The value of the parame ter B was taken 6.7 from [13] and C was set equal 2. In accord with the correlation (4-16) from [13] corresponding reactor to particle diameter ratio is 5.5 for spherical particles, or 11 for Rashig ring shaped particles. The adopted value of the parameter C thus represents the case of relatively strong wall flow which reaches 33% of the total flow rate. The coefficient of radial spread, D, is given by the correlations in paragraph 4.2.2. Its value, however, is not needed at this stage. It is however needed later when converting the results from the dimensionless to dimensional form.

Parameters of reaction: Various orders of the reaction were used in the calculations. The zero-th order was simulated for numerical reasons by $n = 0.1$ and additional values used were $n = 0.5$; 1 and 2. The Damkoehler number was put equal to 0.5 and the adopted values of the activation energy, $\gamma = 8.5$, and the dimensionless adiabatic temperature rise, $U = 0.87$, corresponding to hydrogenation of about 20% solution of an unsaturated solution with a single double bond with the laboratory temperature taken for the reference temperature.

Fig. 6-2 shows the contours of constant concentration in a vertical plane passing through the axis of the reactor. The upper horizontal edge of the frame represents the top of the catalytic layer. The bold arrows show the uniform irrigation of the whole top of the layer. Corresponding contours of constant density of irrigation were shown in Fig. 4-3.

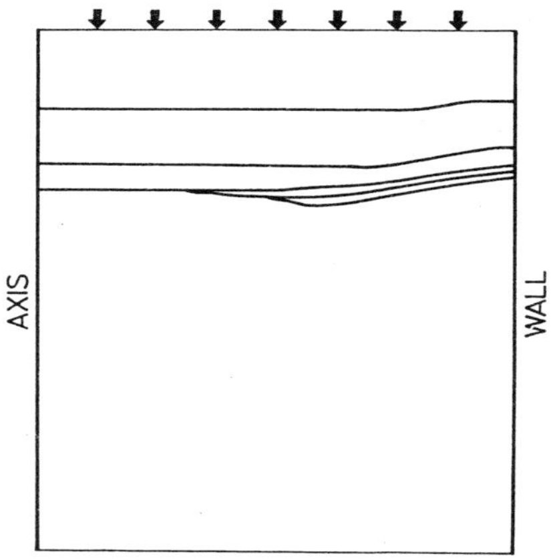

Fig.6-2
Contours of constant concentration in the r-z plane under uniform feed of reacting liquid and $C = 2$, $B = 6.7$, $H = 0.05$, $n = 0.1$, $Da = 0.5$, $\gamma = 8.5$, $U = 0.87$. Contours from top to bottom represent dimensionles concentrations 0.9; 0.7; 0.5; 0.3; 0.1.

The results in Fig. 6-2 were computed for a reaction close to zero order $(n = 0.1)$ and for $H = 0.05$, i.e. for a reactor with relatively shallow layer of the catalyst. In view of the low reaction order the reaction takes place in a narrow reaction zone with a relatively modest curvature of the contours of constant concentration. The shape of the contours of constant concentration reflects the combined effect of the wall flow and increased reaction rate due to the locally diminished density of irrigation in region near the reactor wall. The diminished local density of irrigation near the wall, where the flow over the catalyst surface is depleted of liquid due to the build up of the wall flow manifests itself in that the contours of constant concentration turn initially down toward the bottom. On the other hand, in the immediate vicinity of the wall the reaction rate is somwhat lower as the liquid here is diluted by the wall flow and no reaction takes place directly in the wall flow. The decrease of the reaction rate then manifests itself by turning the direction of the contours upwards toward the top of the bed.

However, the decelerating effect of the wall flow manifests itself substantially more strongly in the bottom part of Fig. 6-2. The last lowest contour in this figure is the contour of the dimensionless concentration 0.1 and the calculation was terminated when the mean concentration dropped just below 0.01. It is thus obvious that owing to the existence of the wall flow one needs more than doubled depth of the bed in order to increase conversion from 90% to 99%.

The isotherms corresponding to the situation in Fig. 6-2 will not be shown as they are complementary to the contours of constant concentration thanks to the invariant in Eq. (6-26).

Fig. 6-3 depicts contours of constant concentration again for a nearly zero order reaction ($n = 0.1$) but with the feed of liquid in the form of a narrow jet ($r_1 = 0.1$) located in the axis of the reactor (see the bold arrows in the figure). The contours of constant density of irrigation for this case were shown in Fig. 4-4. The numerical value of $H = 0.0456$ again indicates a relatively shallow catalyst bed.

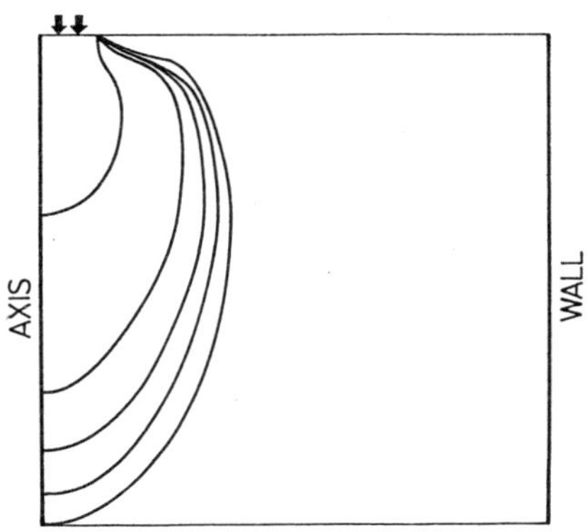

Fig.6-3
Contours of constant concentration in the r-z plane in a bed fed by a narrow nozzle ($r_1 = 0.1$) and $C = 2$, $B = 6.7$, $H = 0.0456$, $n = 0.1$, $Da = 0.5$, $\gamma = 8.5$, $U = 0.87$. Contours from left to right represent dimensionles concentrations 0.9; 0.7; 0.5; 0.3; 0.1.

The reaction zone is narrow but turned almost into a vertical position. With this distributor the maximum density of irrigation reaches as much as 100 times the mean density of irrigation of the catalyst. In the shallow bed with so much axially concentrated flow almost no liquid reaches the wall and hence almost no wall flow forms. On the other hand, the flow even in the bottom part of the reactor is strongly nonuniform.

Due to this type of liquid distribution the catalytic reaction takes place at effectively higher densities of irrigation. In spite of this, the required depth of bed for 99% conversion is somewhat less than in the previous case (lower H) thanks to the elimination of the wall flow. A significant difference between the central jet and the uniform distributor, apparent from Figs 6-2 and 6-3 is that the central jet exposes (also thermally) more the central core of the reactor, while the uniform distributor exposes the entrance section.

Analogous calculations, carried out for higher reaction orders indicate that due to the increasing dependence of the reaction rate on the degree of conversion the depths of the catalyst required for 99% conversion are higher and higher. For the reaction with $n = 0.5$ the use of disc distributors with smaller and smaller radius r_1 leads to a marked decrease of the depth of the catalyst required for 99% conversion. This led to the conclusion that the adverse effect of the wall flow causing lower net reaction rates is in this case more significant than the decrease of the net reaction rate due to increased local densities of irrigation.

For a first order reaction the necessary depth of the catalyst is even higher and this fact considerably reduces the effect of the region immediately below the distributor with its peculiarities of flow distribution. Even for narrow jet distributors the wall flow reaches near the supporting grid significant magnitudes and even the very narrow jet $(r_1 = 0.1)$ provides for a sensible reduction of the depth of the bed required for 99% conversion in comparison with the uniform feed.

For a second order reaction $(n = 2)$ the effect of the entrance region below the mouth of the feed distributor is already very weak and the contours of constant concentration for different distributors are almost the same. The dominating effect is that of the wall flow manifesting itself by slanting contours of constant concentration in region near the wall toward the supporting grid of the reactor.

Apart from the field of the concentration and temperature the solution obtained also gave the mean concentrations. Let us define the integral mean concentration of the reactant, c_m, as

$$c_m = \int_0^1 2f\, c\, r\, dr + c_w W \tag{6-33}$$

c_m is thus the concentration given rise to by perfect mixing (cup mixing concentration) of the outlet stream of liquid including the wall flow.

Fig. 6-4 depicts the dependence of the mean concentration of the reactant as a function of the dimensionless depth of the catalyst bed, expressed by the product xH, for a nearly zero order $(n = 0.1)$ reaction. The curve 4 in this figure represents the ideal case of a uniformly irrigated reactor free of the effect of the wall flow $(C = \infty)$. A comparison with a uniformly irrigated reactor under the strong effect of the wall flow $(C = 2)$, see curve 1, indicates that due to the local decrease of the density of irrigation brought, about by the formation of the wall flow, the reaction proceeds faster. Provided that conversions below about 90% are sufficient, the existence of the wall flow causes a certain reduction of the necessary legth of the catalyst bed.

The conclusion, however, is markedly different if high degrees of conversion are required. In this case the required depths of the bed are substantially greater due to the existence of the wall flow. The curves 2 and 3 were computed for the case of a disc distributor $(r_1 = 0.5)$ and a narrow jet $(r_1 = 0.1)$. In both these cases the reaction is sufficiently fast to reach high degrees of conversion before the appearance of any appreciable amount of the wall flow. In comparison with the uniform irrigation, however, the required depth of the catalyst layer, excepting very low conversions, is mostly higher. In comparison with the uniform irrigation, the more uniformly distributed reaction rate along the catalyst depth may appear as advantageous and therefore more uniform temperature strain of the catalyst.

The analogous dependence of the outlet cup mixing mean concentration on the dimensionless depth of the bed is plotted in Fig. 6-5 for a reaction of $n = 0.5$ order. The curves 1-4 have the same meaning as on the previous figure. The mutual position of individual curves is also similar as before. The wall flow at uniform irrigation of the bed (curve 1) initially causes an increase of the net reaction rate by transferring part of the liquid into the wall flow, which in turn causes faster reaction at lower local irrigation rates. For this reason the curve 1 appears for lower conversions above the curve 4 representing the idealized situation of the uniform irrgation and the absence of the wall flow. For higher conversions, however, this curve lies above curve 4, because the relatively slow exchange of liquid between the flow over the catalyst surface near the wall and the wall flow is the rate limiting step. In this region the existence of the wall flow, similarly as before, causes a significant increase of the necesssary depth of the bed for a given conversion.

Position of the curves 2 and 3, corresponding to the disc $(r_1 = 0.5)$ and narrow jet $(r_1 = 0.1)$ distributors, is shifted to the left of the curves 1 and 4. This indicates that in shorter reactors (compare value of xH in Figs 6-4 and 6-5) the lower net reaction rate due to the accummulation of the liquid flow into the axial

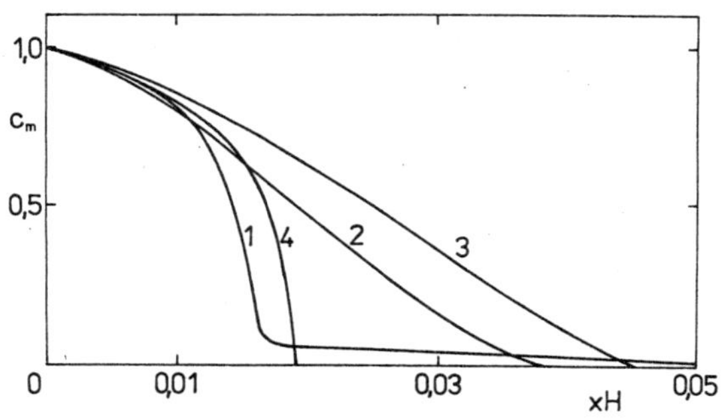

Fig.6-4
Cup mixing mean outlet concentration as a function of dimensionless bed depth for $n = 0.1$, $B = 6.7$, $Da = 0.5$, $\gamma = 8.5$, $U = 0.87$. Curve 1 - uniform feed of reacting liquid, $C = 2$; 2 - $r_1 = 0.5$, $C = 2$; 3 - $r_1 = 0.1$, $C = 2$; 4 - uniform feed of reacting liquid and no wall flow effect $(C = \infty)$.

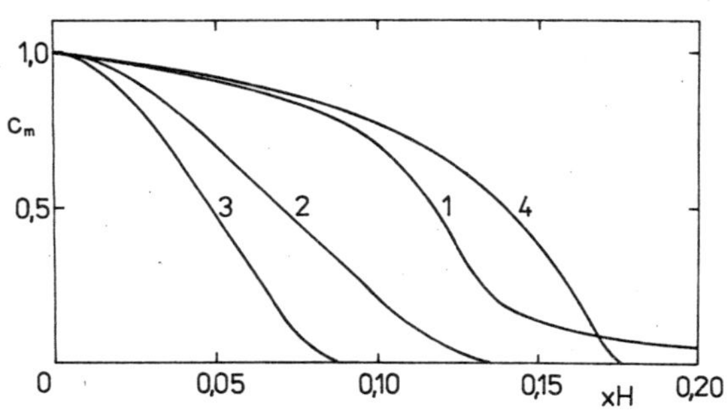

Fig.6-5
Cup mixing mean outlet concentration as a function of dimensionless bed depth for $n = 0.5$, $B = 6.7$, $Da = 0.5$, $g = 8.5$, $U = 0.87$. Curve 1 - uniform feed of reacting liquid, $C = 2$; 2 - $r_1 = 0.5$, $C = 2$; 3 - $r_1 = 0.1$, $C = 2$; 4 - uniform feed of reacting liquid and no wall flow effect $(C = \infty)$.

region is less significant than the increase of the net reaction rate due to restriction of the effect of the wall flow.

Similar plots constructed for higher order reactions exhibit a gradual diminishing effect of the entrance region with strong flow nonuniformities due to the nonuniform feed. For instance, for a second order reaction the curves 1 and 4 are very close to each other and their course reflects essentially only the effect of the wall flow. Even this effect, however, is relatively weaker, since the reduction of the net reaction rate by the slow exchange of liquid with the wall flow is no longer so dramatic.

The decrease of the net reaction rate due to the effect of the wall flow at high reaction rates and low reaction order may be estimated on the basis of the following consideration:

At high reaction rates the concentration of reactant in the liquid trickling down the surface of the catalyst is close to zero $(c = 0)$. If we assume that from the viewpoint of flow distribution the reactor is close to the dynamic equilibrium, the axial variations of the wall flow are negligible $(\mathrm{d}W/\mathrm{d}x = 0)$. Substituting these assumptions into Eq. (6-22) we obtain after a small rearrangement the following expression

$$\frac{\mathrm{d}(\ln c_{\mathrm{w}})}{\mathrm{d}(xH)} = -2BC \tag{6-34}$$

from which it follows that the decrease of the concentration obeys approximately first order equation with the constant $1/(2BC)$. The concentration of the reactant thus decreases by approximately 63% from the original value over the depth of the bed equal $1/(2BC)$ dimensionless length units. The overall "reaction rate" under these conditions is therefore higher in reactors with smaller particles.

To summarize, the effect of flow inhomogeneity in catalytic trickle bed reactors may be divided into two categories:

a) The effect of the entrance region under the mouth of the liquid distributor with marked flow inhomogeneities due to eventual nonuniform feed and the formation of the wall flow.

b) The effect of the wall flow in region where more or less dynamic equilibrium of the flow prevails.

On the basis of the solution of the pseudohomogeneous radially distributed model with the wall flow one can characterize these two categories more specifically as:

a) The effect of flow inhomogeneities in the entrance region of the reactor becomes important particularly for low order reactions and if conditions of the feed are such that the reaction may already start in the entrance region. Due to these flow inhomogeneities markedly curved reaction zones form whose extent is usually greater than the extent of the ideal horizontal reaction zones. Greater extent of the reaction zones may bring about a modest increase of the net reaction rate. The highly nonuniform flow and high local irrigation rates lead to generally lowered net reaction rate. The decrease of the local density of irrigation due to the formation of the wall flow causes an increase of the local reaction rates and may lead to an interim increase of the net reaction rate, mostly, however, only at low conversions.

b) The wall flow causes that part of the liquid flows down a noncatalytic surface and reacts only gradually as a consequence of the dynamic exchange of liquid between the wall flow and the flow through the catalyst bed. In the region of dynamic flow equilibrium this phenomenon always causes the net reaction rate to decrease and the more so the lower the reaction order.

6.2. THE EFFECT OF FLOW INHOMOGENEITIES OF TRICKLING LIQUID ON THE BEHAVIOUR OF A COOLED TRICKLE BED CATALYTIC REACTOR IN THE PRESENCE OF PARALLEL HEAT CONDUCTION IN THE SOLID PHASE

An important assumption of the model analyzed in the previous paragraphs of this chapter was the assumption 3) according to which the only mechanism responsible for the radial transport of heat and mass was the radial dispersion of the liquid itself. This assumption could be supported by, for instance, papers of Hashimoto and Muroyama [69, 70] who found the radial Peclet number for heat and mass transfer to be essentially the same.

Other results [71, 321], including our own [30], however, indicate that among the possible mechanisms of radial heat transport the contribution of the thermal conductivity of particles themselves is not negligible. The importance of the conductive contribution is relatively higher, particularly at low irrigation densities and under low gas velocities. The material of the particles, or their thermal conductivity also has a significant effect.

Since in catalytic trickle bed reactors the mean density of irrigation is usually very low, by as much as an order of magnitude in comparison with the absorption columns, the heat transport via conduction through the particles represents a parallel mechanism to the dispersion with the trickling liquid.

This situation is shown schematically in Fig. 6-6 depicting a part of the catalyst layer in the proximity of the wall. The arrow A symbolizes the heat transfer together with the liquid rivulet trickling down the surface of the catalyst, shown in the figure as a dark streak. The arrow B symbolizes the heat transport by conduction via the material of the particles and the points of their contact. Both mechanisms, shown in Fig. 6-6 separately, however, work simultaneously also in places irrigated by the liquid.

Two C arrows symbolize an analogous situation during heat transfer into the wall mediated by the wall flow, or by the exchange of liquid between the wall flow and the flow down the catalyst surface. The arrow D, on the other hand, represents direct conductive flux of heat into the wall mediated by the thermal conductivity of the catalyst particles and their contact with the wall.

In accord with Fig. 6-6 we shall expand the formulation of the problem also to cases of nonadiabatic reactors, i.e. reactors externally cooled by a cooling medium of temperature T_c.

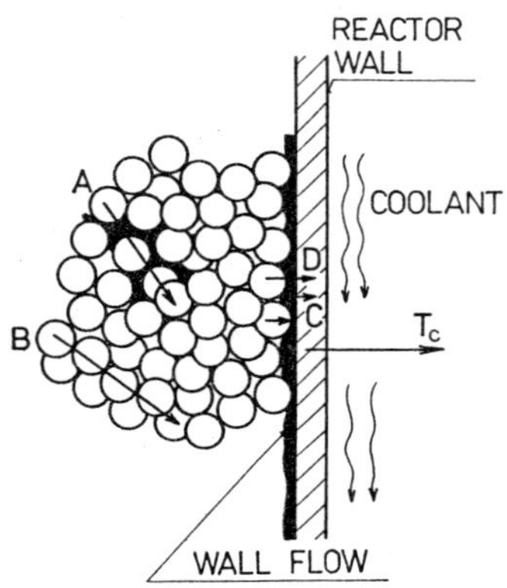

Fig.6-6
Sketch of the mechanisms of heat transfer in a trickle bed reactor.

At low mean densities of irrigation an important role may play also axial convection with the streaming gas, mostly hydrogen. The cause for the relative significance of this transport is the same as before.

All modifications of the model, i.e. its extension by parallel conduction in the radial direction via the particles, axial convection via the streaming gas and cooling the reactor wall, concern the enthalpy balances. The material balances on the liquid as a whole, Eqs (4-5), (4-6), as well as the reactant, Eqs (6-14), (6-16), remain without change. Eq. (6-16) may be still rearranged formally to a simpler form using Eq. (6-4) to obtain

$$-f\frac{\partial c}{\partial r} = BCW(c - c_w) \quad \text{for} \quad r = 1 \tag{6-35}$$

Dispersion of heat in the bed of catalyst in the radial direction together with the trickling liquid is expressed by the second term on the left hand side of Eq. (6-9). The effective conductive heat flux in the radial direction may be expressed by means of the effective radial thermal conductivity, K_R, and the expression

$$-K_R\frac{\partial T'}{\partial r} \tag{6-36}$$

If we rationally assume that the transfer of heat between the gas, on the one hand, and the catalyst and the liquid on the other hand, is very intensive (the volume heat transfer coefficient approaches infinity), the temperature of the gas will equal T'. Axial convective heat transfer together with the flowing gas may then be expressed by the following term

$$T'V\varrho_g c_{pg} \tag{6-37}$$

where V designates the superficial gas velocity while ϱ_g and c_{pg} are its density and specififc heat capacity.

Let us define for brevity still two following quantities

$$H_R = \frac{hK_R}{r^2 \varrho c f} \tag{6-38}$$

and

$$P = \frac{p\,\varrho_g c_{pg}}{f_0 \gamma c_p} \tag{6-39}$$

The dimensionless quantity P designates the ratio of the heat capacity fluxes in the gas and the liquid phase while the quantity H_R may be expressed as a product of the simplexes (h/r_0), (d_p/r_0) and the quantity P divided by the Peclet number for the effective thermal conductivity of the particles, $Vc_{pg}\varrho_g d_p/K_R$.

The enthalpy balance of the type of Eq. (6-9) incorporating the above modifications then takes the following form

$$-\frac{\partial(fT)}{\partial x} - P\frac{\partial T}{\partial x} + H\frac{1}{r}\frac{\partial}{\partial r}\left(r\frac{\partial(fT)}{\partial r}\right) + H_R\frac{1}{r}\frac{\partial}{\partial r}\left(r\frac{\partial T}{\partial r}\right) + U\gamma\, Da_n\frac{R}{R_0} = 0 \tag{6-40}$$

in which we used for the time being the more general form of the reaction rate term.

Upon separating the variables and changing over to the explicit form for an n-th order reaction we obtain

$$(f + P)\frac{\partial T}{\partial x} - H\left[f\frac{\partial^2 T}{\partial r^2} + 2\frac{\partial f}{\partial r}\frac{\partial T}{\partial r} + \frac{1}{r}f\frac{\partial T}{\partial r}\right] - \\ - H_R\left[\frac{\partial^2 T}{\partial r^2} + \frac{1}{r}\frac{\partial T}{\partial r}\right] U\gamma\, Da_n c^n \exp\left[\frac{T}{1 + T/\gamma}\right] = 0 \tag{6-41}$$

The fact that the heat is transported by effective conduction also by the catalyst particles will show also in the boundary condition for Eq. (6-41). Namely,

$$-\left(f + \frac{H_R}{H}\right)\frac{\partial T}{\partial r} - T\frac{\partial f}{\partial r} = B(fT - CWT_w) \text{ for } r = 1 \tag{6-42}$$

With the aid of the boundary condition (6-4) the equation (6-42) may be transformed to the following, formally simpler, form

$$-\left(f + \frac{H_R}{H}\right)\frac{\partial T}{\partial r} = BCW(T - T_w) \text{ for } r = 1 \tag{6-43}$$

In the enthalpy balance in the wall flow will also appear the transport of heat into the coolant. The temperature of the coolant is regarded constant and the heat flux is expressed by means of the coefficient h_w.

Upon introducing the dimensionless heat transfer coefficient into the wall as

$$H_w = \frac{h_w h}{r_0 \varrho c_p f_0} \tag{6-44}$$

the appropriate heat balance may be written as

$$\frac{d(WT_w)}{dx} + 2H_w(T_w - T_c) + 2H\frac{\partial(fT)}{\partial r} + 2H_R\frac{\partial T}{\partial r} = 0 \text{ for } r = 1 \tag{6-45}$$

By separation of the dependent variables and making use of the boundary condition (6-44) one obtains two alternative forms of the heat balance in the wall flow. Namely

$$W\frac{dT_w}{dx} + 2(Hf + H_R)\frac{\partial T}{\partial r} + \frac{dW}{dx}(T_w - T) + 2H_w(T_w - T_c) = 0$$

$$\text{for } r = 1 \tag{6-46}$$

or

$$W\frac{dT_w}{dx} - 2HBCW(T - T_w) + \frac{dW}{dx}(T_w - T) + 2H_w(T_w - T_c) = 0$$

$$\text{for } r = 1 \tag{6-47}$$

One could write two more alternative forms where the derivative of the wall flow is replaced by the gradient of the density of irrigation in the radial direction at the point $r = 1$.

For a better understanding of the model and testing of internal consistency of the numerical solutions it is useful to formulate still macroscopic balances of heat and mass. Let us integrate first the enthalpy balance (6-40) over the radius with the weight $2r$ in the limits $0 - 1$. After some arrangement of the third and the fourth term using the boundary condition (6-42) and the heat balance in the wall flow, Eq. (6-46), we obtain the following equation

$$-\frac{\partial}{\partial x}\int_0^1 (f + P)\, T\, 2r\, \mathrm{d}r - \frac{\mathrm{d}(WT_{\mathrm{w}})}{\mathrm{d}x} - 2H_{\mathrm{w}}(T_{\mathrm{w}} - T_{\mathrm{c}}) + U\,\gamma\, Da_n \int_0^1 2r\frac{R}{R_0}\,\mathrm{d}r = 0 \tag{6-48}$$

In words this equation says that the heat produced by the reaction over an infinitesimal length of the reactor (4-th term) is consumed to heat the liquid and the gas stream (1-st term), to heat the wall flow (2-nd term) and the rest is taken by the coolant (3-rd term).

Additional integration over the axial coordinate in the limits 0 - 1 leads to the following equation

$$- \int_0^1 (f + P)\, T_{x=1} 2r\, \mathrm{d}r - (WT_{\mathrm{w}})_{x=1} - 2H_{\mathrm{w}} \int_0^1 (T_{\mathrm{w}} - T_{\mathrm{c}})\, \mathrm{d}x + \\ + U\gamma\, Da_n \int_0^1 \mathrm{d}x \int_0^1 2r\frac{R}{R_0}\,\mathrm{d}r = 0 \tag{6-49}$$

expressing that the heat liberated in the whole reactor by the reaction (4-th term) is consumed to heat the outlet liquid and gas stream (1-st term), to heat the outlet wall flow (2-nd term) and the rest is carried away by the coolant (3-rd term).

The macroscopic balance of the reacting species obtained by integration of Eq. (6-5) over the whole reactor volume

$$Da_n \int_0^1 \mathrm{d}x \int_0^1 2r\frac{R}{R_0}\,\mathrm{d}r = 1 - \int_0^1 2r\,(fc)_{x=1}\,\mathrm{d}r - (Wc_{\mathrm{w}})_{x=1} = 1 - c_{\mathrm{m}} \tag{6-50}$$

relates the mean reaction rate in the whole reactor and the mean outlet cup mixing concentration.

Upon introducing the last result into Eq. (6-49) we obtain after some arrangement the following equation

$$U\gamma = U\gamma\, c_{\mathrm{m}} + T_{\mathrm{m}} + P\int_0^1 T_{x=1}\, 2r\,\mathrm{d}r + 2H_{\mathrm{w}}\int_0^1 (T_{\mathrm{w}} - T_{\mathrm{c}})\,\mathrm{d}x \qquad (6\text{-}51)$$

in which the mean cup mixing temperature is defined as

$$T_{\mathrm{m}} = \int_0^1 f\,T\,2r\,\mathrm{d}r + WT_{\mathrm{w}} \qquad (6\text{-}52)$$

i.e. analogously to the the mean concentration. Clearly, both c_{m} and T_{m} are taken at the point $x = 1$.

From Eq. (6-51) and the foregoing macroscopic balances it follows that the maximum dimensionless local temperature, equally as in the adiabatic reactor, cannot exceed $U\gamma$ and under cocurrent flow in a cooled reactor will not even reach this value.

Under parallel conduction of heat via the catalyst pellets the local temperatures and concentrations do not obey the invariants (6-26) and (6-32). Only under the adiabatic mode of operation of the reactor $(H_{\mathrm{w}} = 0)$ and zero flow rate of gas $(P = 0)$ one can consider Eq. (6-51) as an analog of the mentioned invariants valid, of course, only for the integral mean quantities c_{m}, T_{m}.

6.3. THE EFFECT OF FLOW INHOMOGENEITIES OF TRICKLING LIQUID ON CATALYTIC HYDROGENATION OF CYCLOHEXENE IN AN ADIABATIC CATALYTIC REACTOR

Catalytic reactions carried out in trickle bed reactors (hydrogenations, oxidations) are typically strongly exothermic. A study of flow inhomogeneities in trickle bed reactors using hydrogenation of cyclohexene as a model reaction appears to have certain advantages. Hydrogenation of cyclohexene to cyclohexane takes place in the liquid phase at sufficient reaction rate at normal pressure and moderate temperatures. It is a reaction in many respects typical for the whole class of hydrogenations with the reaction heat (see Table 6-1) exceeding 30 kcal per mole of double bonds.

On the basis of experimental studies [65] the rate of hydrogenation of cyclohexene in the liquid phase and 1 atmosphere may be described by the following kinetic equation

$$R = 0.00223\,(c')^{0.1}\exp[-2516/T']\,X_{\mathrm{H}}^{0.61} \qquad (6\text{-}53)$$

$$(R[\text{mole/cm}^3\text{s}];\ c'[\text{mole/liter}])$$

From this equation it follows that we are dealing with a nearly zero order reaction with respect to the substrate (0.1-th order was selected for numerical reasons of subsequent solutions of the model equations and the deviations from perfectly zero order model are entirely negligible) and 0.61-th order with respect to hydrogen. The mole fraction of hydrogen in the gas phase is designated by X_H.

Table 6-1

Quantity	Numerical value
thermodynamic reaction heat	142 kJ/mole of cyclohexene (34 kcal/mole)
reaction heat, (ΔH)	28.4 kJ/mole of feed (at c_0)
density of liquid, ϱ	0.78 g/cm^3 (at T_0)
surface tension of liquid	25 dyn/cm (at T_0)
specific heat of liquid, c_p	151 kJ/mole K (36 kcal/mole K)
inlet temperature, T_0	303 K
inlet concentration, c_0	1.86 mole/liter
dimensionless adiabatic temperature rise	0.1248 (at c_0, T_0, $(-\Delta H)$)
dimensionless activation energy, γ	8.3 (at T_0)
Damkoehler number, Da	0.9 (at $f_0 = 0.00982$ cm/s) 1.8 (at $f_0 = 0.00491$ cm/s)

The reaction rate, expressed in dimensionless quantities, takes the following form

$$\frac{R}{R_0} = c^{0.1} \exp\left[\frac{T}{1 + T/\gamma}\right] X_H^{0.61} \tag{6-54}$$

with that the reference reaction rate, R_0, appearing also in the definition of the Damkoehler number (6-17) (subscript n omitted for brevity), is the reaction rate at temperature T_0, concentration c_0 and the mole fraction of hydrogenin the gas phase equal unity, $X_H = 1$.

In the course of the reaction, however, the mole fraction of hydrogen is reduced by the vapours of the reacting cyclohexene/cyclohexane mixture. The instantaneous mole fraction of hydrogen in the gas phase therefore depends on temperature and composition of the phases. Since, however, the thermodynamic properties of the reactant and product are very close the dependence on composition may be safely neglected. The dependence of the mole fraction of hydrogen in the gas phase at total pressure of 1 atm may then be expressed by the following expressionin terms of the dimensionless variables

$$X_H = 1 - 27496 \exp\left[-\frac{3618.5}{T_0}\frac{1}{1 + T/\gamma}\right] \tag{6-55}$$

in which we used the Clausius-Clapeyron expression for the temperature dependence of the vapour pressure of cyclohexane.

Solution of the model of the catalytic hydrogenation was then carried out using the equations derived in the previous paragraph with the reaction term given by Eq. (6-54) which differs from the form in Eq. (6-14) only in the coefficient expressing the dependence of the reaction rateon the mole fraction of hydrogen in the gasphase. Its relationship to the dimensionless temperature is given by Eq. (6-55).

Solution of the model was carried out for two types of the feed of the substrate into the reactor. In the first case the feed was carried out by means of the central jet of radius 0.15 r_0. The initial condition then reads

$$f = (1/0.15)^2 \quad \text{for} \quad 0 \leqq r < 0.15 \quad \text{and for} \quad x = 0$$

$$f = 0 \quad \text{for} \quad 0.15 < r \leqq 1 \text{ and for} \quad x = 0 \tag{6-56}$$

In the second case the substrate was fed by the same jet as before, but there was a 50 mm deep layer of inactive glass spheres 2 mm in diameter placed on top of the catalyst layer. In this case an analytical solution of Eq. (4-5) was carried out also for the catalytically inactive layer and the obtained result of flow distribution represented the initial flow distribution for the solution within the catalyst bed itself.

Necessary values of the coefficient of radial spread for the catalyst particles were computed from the correlation of Onda [54], incoporating also the effect of surface tension of the trickling liquid, Eq. (4-20). The coefficient C, characterizing the magnitude of the wall flow effect, was computed from the correlation (4-44) with the result $C = 1.204$. Additional data necessary for the formulation of the problem, are summarized in [26].

The experimental runs of hydrogenation of cyclohexene in cyclohexane were carried out in a reactor 3 cm in diameter packed with the catalyst containing 3% of Pd on activated carbon (Supersorbon H8-3 Degussa) with a commercial designation CHEROX 41-00. The particles were cylindrical pellets 0.35 cm in diameter, 0.5 cm high (equivalent particle diameter $d_p = 0.45$). The height of the catalyst bed was varied between 2 and 28 cm (reference height $h = 28$ cm).

A layer of fresh catalyst was prepared for each run and the hydrogenation was carried out for two mean densities of irrigation (see Table 6-1) and for two types of initial liquid distributions as described above. The reactor was a glass tube thermally insulated by polystyrene foam. The feed liquid was a 20% solution of cyclohexene in cyclohexane but it was necessary before each experimental run to irrigate the fresh catalyst bed by pure cyclohexane with the aim to liberate and remove the large amount of the heat of wetting.

The steady state mean outlet concentration of cyclohexane was measured monitoring at the same time the steady state by a set of thermocouples placed in the bed.

Fig. 6-7 plots the mean outlet concentration, c_m, defined in Eq. (6-33), in dependence on the dimensionless axial coordinate. Graphical symbols distinguish between the two types of liquid feed (jet, or jet with inactive layer). The results in Fig. 6-7 were obtained for the higher of the two densities of irrigation ($Da = 0.9$). The agreement of the experimental results with the computed curves is satisfactory if we consider that each point represents a freshly prepared bed and that the curves were not optimized with respect to parameters.

The curves 1 and 2 in the figure correspond to direct irrigation and irrigation in the presence of the inactive layer. Mutually they differ little owing to the slow rate of conversion at higher densities of irrigation. Both curves, however, differ in region of higher conversions from curve 3 computed for the idealized case of the reactor free of the wall effect.

The following Fig. 6-8 depicts analogous results obtained for a lower mean density of irrigation ($Da = 1.8$). Both computed dependences markedly differ in the region of higher conversions due to the slow rate of liberation of the unreacted cyclohexene from the wall flow. The slower net reaction rate is therefore more marked for indirect irrigation of the catalyst bed with an inactive

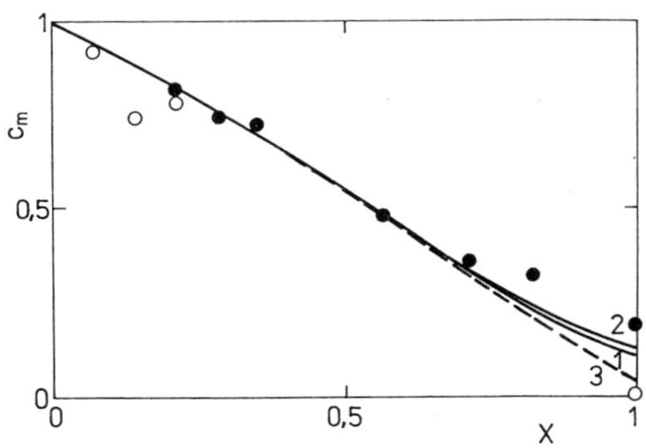

Fig.6-7
Cup mixing mean outlet concentration as a function of the bed depth for hydrogenation of of cyclohexene on Pd catalyst for $f_0 = 0.00982$ cm/s (see Table 6-1). Empty circles - experiments with direct irrigation of the bed by a narrow jet; Full circles - experiments for indirect irrigation of the catalyst bed mediated by a layer of inert glass spheres. Curve 1 - calculated for directly irrigated bed; 2 - calculated for indirectly irrigated bed; 3- calculated for wall flow free ($C = \infty$) reactor.

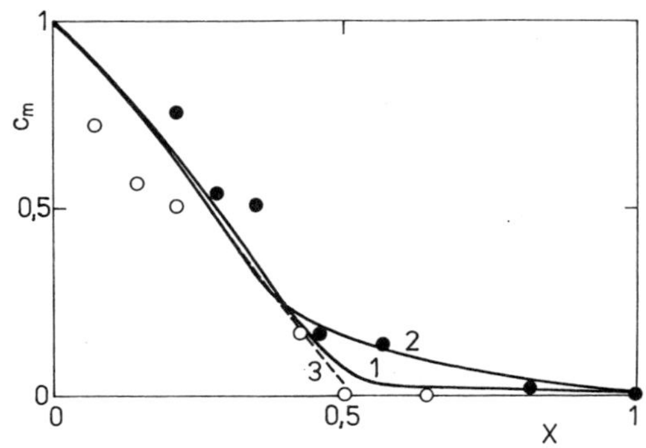

Fig.6-8
Cup mixing mean outlet concentration as a function of the bed depth for hydrogenation of cyclohexene on Pd catalyst for $f_0 = 0.00491$ cm/s (see Table 6-1). Empty circles - experiments with direct irrigation of the bed by a narrow jet; Full circles - experiments for indirect irrigation of the catalyst bed mediated by a layer of inert glass spheres. Curve 1 - calculated for directly irrigated bed; 2 - calculated for indirectly irrigated bed; 3- calculated for wall flow free ($C = \infty$) reactor.

layer of glass spheres. This leads to faster formation of the wall flow. The curve 3 computed for comparison for the reactor free of the wall effect does not at all foresee this course of the dependence of the outlet concentration while the experimental points confirm this "tailing", predicted by the pseudohomogeneous radially distributed model, at least within experimental error.

As has been shown the solution of the model also provides the field of concentration of cyclohexene within the catalyst bed. Fig. 6-9 shows the contours of constant concentration under direct irrigation of the bed by the jet for the lower of the two mean densities of irrigation. This case was globally characterized by curve 1 in Fig. 6-8. Fig. 6-9 shows that under irrigation by a narrow jet the flow is markedly axially concentrated. In the region around the reactor axis the density of irrigation considerably exceeds the mean and the reaction proceeds slower. Only a minor cuvature of the lowest of the shown contour $(c = 0.1)$ near the wall suggests the impeding effect of the wall flow. From this it follows that the slower net reaction rate in region of high conversions was in case of the jet distributor the combined result of the axially concentrated flow and the wall flow.

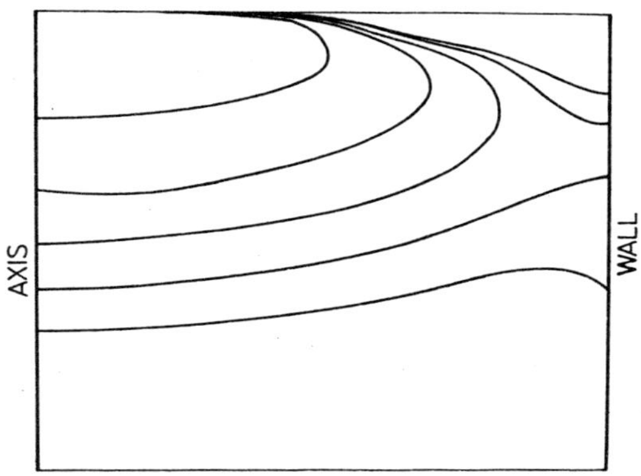

Fig.6-9
Computed contours of constant concentration in the r-z plane of the reactor for hydrogenation of cyclohexene on Pd catalyst and $f = 0.00491$ cm/s. Contours from the top left hand corner to bottom represent dimensionless concentration 0.9, 0.7, 0.5, 0.3, 0.1. Contours 0.1 and 0.3 have an upper and lower branch.

Fig. 6-10 again depicts contours of constant concentration for the reactor irrigated indirectly across the layer of inactive particles and again for the lower of the two densities of irrigation . The curve 2 in Fig. 6-8 gives the corresponding dependence of the mean outlet concentration. From Fig. 6-10 it is apparent that under these conditions the hydrogenation of cyclohexene takes place almost exclusively under the impeding effect of the wall flow. This is shown, on the one hand, in the characteristic sloping of the contours of constant concentration and, on the other hand, by the relatively large extent of the "empty" area below the last contour. The last contour corresponds to $c = 0.1$ and the "height" of the empty area represents the length of the bed required for a 99% conversion. The dominance of the wall flow effect reflects the slow release of cyclohexene, trapped in the wall flow, into the flow over the catalyst surface where its concentration is substantially lower owing to the relatively high reaction rate.

The application of the model of heterogeneous catalytic hydrogenation of cyclohexene in a trickle bed reactor represents a useful tool for a more accurate prediction of the conversion, particularly if a high degree of conversion is required and further for correct interpretation of the cause of the appearance of peculiarities on the conversion curves in case of processes with a strong interaction of heat transfer, mass transfer, flow distribution and chemical reaction.

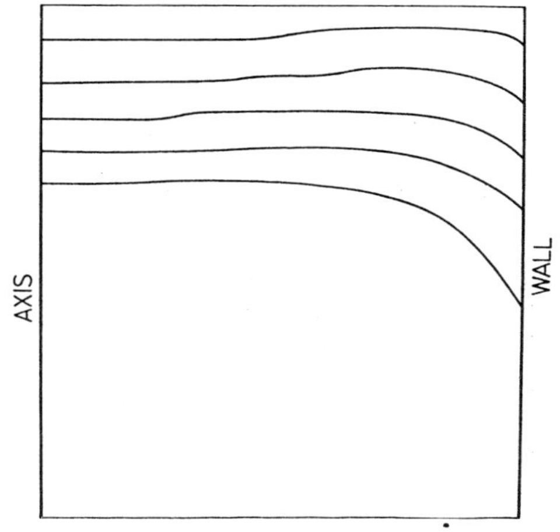

Fig.6-10
Computed contours of constant concentration in the r-z plane of the reactor for hydrogenation of cyclohexene on Pd catalyst at indirect irrigation of the bed mediated by an inert layer of glass spheres and $f_0 = 0.00491$ cm/s. Contours from the top to bottom represent dimensionless concentration 0.9, 0.7, 0.5, 0.3, 0.1.

6.4. THE RELATION BETWEEN FLOW INHOMOGENEITIES OF TRICKLING LIQUID AND PHASE CHANGES IN A HETEROGENEOUS CATALYTIC TRICKLE BED REACTOR

Trickle bed reactors in hydrotreating and hydrocracking operations are often operated under conditions in which the feed reactants, products and inerts are partially evaporated. The resulting distribution of reactants and products between the vapour and the liquid phase masks the intrinsic kinetics and alters the observed activation energy. Altered vapour liquid equilibria have direct consequences for reactor performance and reaction selectivity which are not immediately obvious. LaVopa and Satterfield [324] investigated these phenomena experimentally for two reactions in a laboratory trickle bed reactor and different feed volatilities ranging between all-vapour to largely-liquid. Their experiments as well as the model showed that the greatest effect on conversion of the chang ingliquid feed ratio occurs when liquid volatility is low and when the absolute difference in volatilities between the reactant and the unreacting liquid (solvent) is large. In practice the volatilities of the liquid feed are narrow and the imporant factor is that of the volatilities of intermediates and products. The apparent behaviour of the trickle bed reactor thus strongly depends on the volatilities of the reactant, solvent, products and the gas-to-liquid feed ratio.

However, for the majority of reactions carried out in trickle bed reactors the reaction heats are much larger than the heats of evaporation. For hydrogenation of cyclohexene to cyclohexane, for instance, the reaction heat according to Table 6-1 is 142 kJ/mole of cyclohexene, while the corresponding heat of evaporation is only 30.5 kJ/mole.

Since the heat of evaporation if cyclohexane is almost identical with the value given above for cyclohexene it is apparent that under the adiabatic mode of operation and normal pressure complete local spontaneous evaporation of the reaction mixture by the liberated reaction heat can occur and the same may happen even in fairly dilute reaction mixtures. The reaction kinetics in the gas phase are usually different and undesirable side reactions may be engendered.

Fig. 6-11 shows the temperature dependence of the quantity $\exp\left[\frac{T}{1 + T/\gamma}\right] X_{\mathrm{H}}^{0.61}$, which, as follows from Eqs (6-54) and (6-55), is practically proportional to the rate of hydrogenation of cyclohexene in the liquid phase. The dependence on concentration in the 0.1-th power was adopted for numerical reasons and in practice is insignificant. This dependence is plotted with the numerical value of the activation energy from Table 6-1. Functional dependence of $\exp\left[\frac{T}{1 + T/\gamma}\right] X_{\mathrm{H}}^{0.61}$ on temperature passes through a maximum and subsequently decreases because the increase due to the faster kinetics at higher temperature is offset by the decrease of the partial pressure of hydrogen [65].

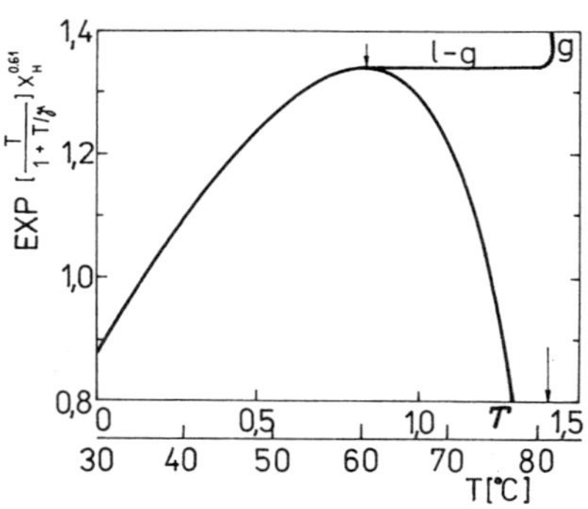

Fig.6-11
Temperature dependence of an expression practically proportional to the rate of liquid phase hydrogenation of cyclohexene. "l-g" designates region of simultaneous gas and liquid phase hydrogenation; "g" represents region of solely gas phase hydrogenation.

Experimentally, however, it was observed that the total rate of hydrogenation of cyclohexene past the maximum on the dependence on temperature does not decrease but remains approximately constant. This course is shown in Fig. 6-11 by the horizontal line "l-g". Here the hydrogenation takes place to an appreciable extent also on the dry surface of the catalyst and the total (liquid and gas phase) rate of hydrogenation is approximately constant.

With an additional increase of temperature the reaction mixture totally evaporates. In this region the hydrogenation takes place solely in the gas phase. The corresponding region in Fig. 6-11 is marked as "g". The evaporation of the reaction mixture takes place generally at a temperature below the boiling point of the mixture, depending on the ratio of the flow rates of the liquid and the gas phase and the rate of gas/liquid interfacial transfer. The arrow in the upper right hand corner of Fig. 6-11 indicates the boiling point of the cyclohexane/cyclohexene mixture.

Transition into the regime of gas phase hydrogenation is accompanied by the change of the kinetics and a marked increase of the reaction rate. The brings along the hazard of overheating and deactivation of the catalyst.

In the previous paragraphs the experimental results analyzed were obtained under conditions in which the reaction took place under all circumstances in the liquid phase. This means that the temperature did not even approach the temperature of the maximum on the curve in Fig. 6-11.

The temperature at which the hydrogenation of cyclohexene is maximal can be determined as the point where $\mathrm{d}(R/R_0)/\mathrm{d}T = 0$. Upon performing the differentiation of Eq. (6-54), substituting from Eq. (6-55), setting the result equal to zero and after some arrangement we obtain for the dimensionless temperature of the maximum, T_{rmax}, of the rate of liquid phase hydrogenation the following expression

$$T_{\mathrm{rmax}} = \frac{3618.5\gamma/T_0}{\ln[27496 + 0.61 \times 27496 \times 3618.5/(\gamma/T_0)]} - \gamma = 333.4\frac{\gamma}{T_0} - \gamma \tag{6-57}$$

The maximum rate of liquid phase hydrogenation of cyclohexane therefore occurs at 333.4 K, i. e. about 60.3°C, or, for conditions given in Table 6-1, at the dimensionless temperature 0.83. The maximum of the reaction rate is marked in Fig. 6-11 by an arrow. After substituting this temperature into Eq. (6-54) we find that the maximum rate of liquid phase hydrogenation equals 1.34 times the rate at inlet conditions.

Since, according to the invariant, the temperature of the reaction mixture can reach at most $U\gamma$, the gas phase hydrogenation or eventual complete evaporation of the reaction mixture may be eliminated by sufficient dilution of the reaction mixture, or, eventually by lowering the temperature of inlet liquid.

In the cooled reactor the invariant (6-26) no longer holds and safe elimination of the evaporation of the reaction mixture cannot be generally ensured by the conditions of the inlet stream. Radial transport of heat in trickle bed reactors is generally not very intensive and, consequently, even a very intensive cooling is no guarantee. Local evaporation is affected by a number of additional parameters [29] such as the properties of the reactor, character of the flow, and catalyst. An estimate of the appearance of local evaporation can be made on the basis of the model of a cooled trickle bed reactor with parallel conduction of heat presented in chapter 6.2.

With this aim we solved the case of hydrogenation of cyclohexene to cyclohexane in a cooled trickle bed reactor in the presence of parallel heat conduction through the catalyst particles themselves. Corresponding balance of the reacting cyclohexene is Eq. (6-14) in which, however, the reaction term was modified to the form in Eq. (6-54). The temperature dependence of the mole fraction of hydrogen in the gas phase is given by Eq. (6-55). The validity of the expression for the reaction rate, however, was restricted below the temperature T_{rmax}, given by Eq. (6-56), at which the hydrogenation takes place at maximum rate. Further course of the temperature dependence of the reaction rate was described by a horizontal straightline ("l-g" in Fig. 6-11). This means that above the temperature T_{rmax} the reaction rate was set equal to the rate at temperature T_{rmax}.

The balance of enthalpy in the trickling liquid was taken in the form (6-41), with the reaction term modified in the same manner as in the material balance. The balance of the reactant in the wall flow remains without change in the form (6-16), while the enthalpy balance in the wall flow (6-46) respects the fact that the reactor is being cooled.

It must be stressed that in the region shown in Fig. 6-11 as "l-g", where the gas phase hydrogenation on dry catalyst pellets contributes significantly to the total rate of hydrogenation , the above balances, strictly speaking, are not valid. This is so because evaporation of the liquid violates the balance on the trickling liquid and its distribution is no longer governed by Eq. (4-5).

From this standpoint one can consider the following results as only estimates of the onset of appearance of hot spots in the reactor where the reaction mixture evaporates by the reaction heat. The course of the reaction once the hot spot has appeared is therefore different from that given by the model.

Qualitatively one can say that with increasing participation of the gas phase hydrogenation (curve "g" in Fig. 6-11) the reaction rate will increase but the temperature rise will weaken by the heat consumed for evaporation of the liquid phase. In addition, it was established experimentally, that gas phase hydrogenation is almost first order with respect to cyclohexene (0.95-th order according to refs [49, 65]). This fact further weakens the progressive increase of the reaction rate.

For the numerical solution of the model of hydrogenation of cyclohexene it was necessary also to determine the parameter H_w, Eq (6-44), characterizing the transfer of heat into the coolant. This quantity was set equal to 10 as the calculation of the thermal resistance for typical conditions showed the resistance to be very small.

Solution of the case of hydrogenation for two identical reactors differing only in the temperature of the feed, T_0, equal to 303 K and 338 K respectively, while the temperature of the coolant, T_c, was in both cases equal to the temperature of the feed, produced in the latter an extensive hot spot. The onset of the zone of evaporation was detected at the distance of about 1/4 to 1/3 of the reactor length from the inlet. The hot spot reached across practically the whole cross section, excepting a narrow ring adhering to the wall. Even at very high intensity of cooling it is thus impossible to sustain isothermal course of hydrogenation and if the temperature of the feed is high it is difficult to supress formation of hot spots with the evaporation of the liquid mixture.

This conclusion confirms also the result of the following calculations carried out for various inlet temperatures. The temperature of the coolant, however, was in all cases 293 K. The increased intensity of cooling did not suffice at the inlet temperature of 343 K to prevent the appearance of the hot spot approximately in the middle of the reactor length reaching about 0.25 r_0 away from the reactor axis and about the same axial dimension. Even a very intensive cooling cannot unambiguously remove the danger of evaporation.

A comparison of two cases of hydrogenation when the temperature of the feed in both cases was 338 K and the temperature of the coolant was 293 K differing in the initial distribution of the liquid feed pointed to significant differences in the course of the reaction: Under uniform initial distribution of liquid and the existence of the wall flow the above conditions of cooling were sufficient to avoid the appearance of hot spots although the radial temperature profiles pointed to marked temperature differences in various parts of the reactor. If, however, the liquid was fed as a narrow jet in the axis, a hot spot of limited extent appeared in the axis of the reactor. Its maximum radial extent was about 0.2 r_0.

This finding is important for the analysis of possible causes of appearance of hot spots in full scale reactors. Although the inhomogeneity of the flow of liquid in the reactor in the given case, on the one hand, enhanced axially concentrated flow, which in turn contributed to the decrease of the net reaction rate, it led, on the other hand, to weakened radial transport of heat. The second of the two effects dominated and ultimately gave rise to the hot spot.

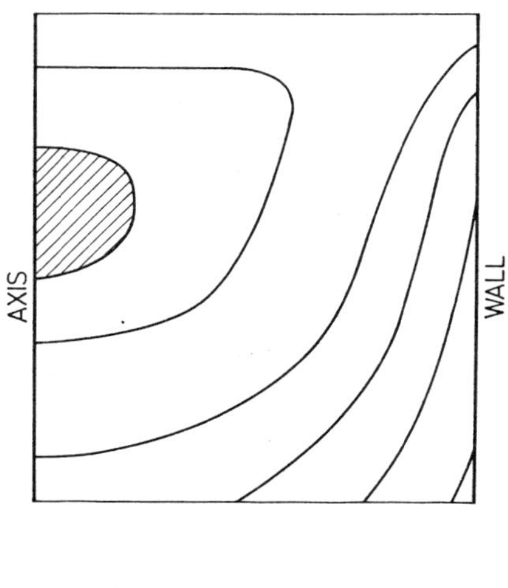

Fig.6-12
Isotherms in the *r-z* plane of the reactor for hydrogenation of cyclohexene at uniform feed of reacting liquid at T =70 C, T =20 C and H =0.1. Shaded area marks region of evaporation of reacting liquid.

Figs 6-12 and 6-13 show the isotherms computed for two reactors with a uniform feed of liquid under the existence of the wall flow. The feed stream had in both cases a temperature of 343 K and the temperature of the coolant was 293 K. Both cases differ in the value of the effective thermal conductivity of the catalyst particles with, however, the same catalytic activity. The differences in the effective thermal conductivities are characterized by values of 0.1 and 0.29 of the dimensionless criterion H_R, defined in Eq. (6-38).

In the case of the lower effective thermal conductivity of the catalyst layer a limited hot spot of liquid evaporation appeared, shown in Fig. 6-12 by the shaded area. The more intensive radial transport of heat due to increased intensity of parallel conduction through the particles themselves, shows in Fig. 6-13 in the generally different character of the isotherms, particularly their lower slope corresponding to more intensive additional heat transport mechanism.

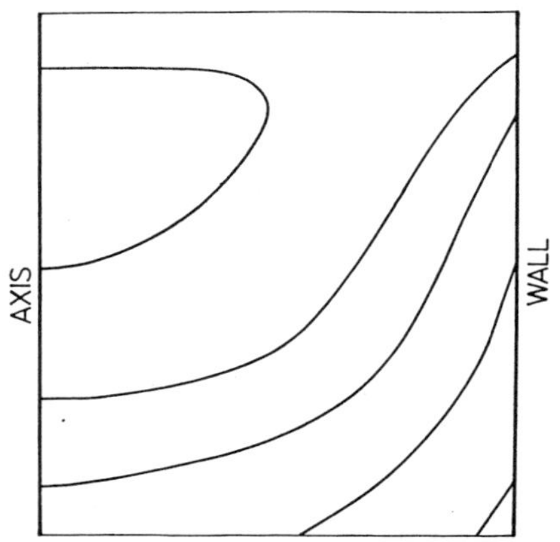

Fig.6-13
Isotherms in the r-z plane of the reactor for hydrogenation of cyclohexene at uniform feed of reacting liquid at T =343K, T =293K and H = 0.29.

A comparison of yet another two computed courses of hydrogenation under the same conditions, i. e. feed temperature 343 K, the temperature of the coolant 293 K and the uniform distribution of the liquid feed into the reactor with the wall flow but different flow rates of gas, also shows markedly different thermal conditions.

The different flow rates of gas were characterized by the value 0.25 and 2.0 of the criterion P, defined in Eq. (6-39). The former of these two cases is identical with the one whose isotherms are shown in Fig. 6-12. The results of the solution for the higher hydrogen flow rate $(P = 2)$ show that the hot spots with evaporation no longer appear. In contrast to Fig. 6-12, however, the isotherms in this case display predominantly axial direction owing to the increased axial convective heat flux.

The last investigated cases suggest a relatively effective method of affecting the thermal regime of the reactor through its hydrodynamics including mode of distributing the liquid feed, flow rate of hydrogen and through modification of the non-catalytic properties of the catalyst including its thermal conductivity.

As already mentioned the previous computational assessment can, in fact, predict correctly only the onset of evaporation. Greater intensities of evaporation violate the formulated heat balances, on the one hand, and the overall mass balances on the trickling liquid (losses of liquid due to evaporation), on the other hand.

A scheme of all phenomena pertaining in a trickle bed reactor, when the reaction takes place both in the liquid and the gas phase, is shown in Fig. 6-14. The top and the bottom line represent the gas and the liquid phase reaction respectively, exhibiting in general different kinetics. The vertical bars represent evaporation and condensation. Both the vertical and horizontal arrows represent phenomena associated with either evolution or absorption of heat.

A mathematical model of hydrogenation of 1,5 cyclooctadiene in a trickle bed reactor accompanied by phase changes as in the above scheme has been presented by Hanika et al [335] and its results have been compared with experimental data from a laboratory reactor. Although the model formulated separate balances for the gas, liquid and solid phases, it partially exhibited the pseudohomogeneous feature in that all phases were taken to be in hydrodynamic and thermodynamic (equal temperatures) equilibrium. Since the model did not consider maldistribution of the flow of phases, nor the existence of the wall flow, it will not be presented here. In the context of this book, however, it is worth noting that eventual success or failure of the model to describe the experimental data depended strongly on the correlation for the degree of wetting of the catalyst surface used in the model equations. This factor is thus believed to be crucial for the modelling of trickle bed reactors including the effect of flow inhomogeneities.

The importance of the degree of wetting in trickle bed reactors has been confirmed also by Sicardi et al [317, 319], Baldi et al [322] and others. Sicardi et al [317] stress also the effect of the quality of wetting (stagnant or semistagnant zones) and recommend dynamic methods, see e.g. Mills and Duduković [318], of evaluating it.

The existence of dry zones on the catalyst surface may in general either increase or decrease the observed reaction rate. Sakornwimon and Sylvester [320] show that if the limiting reactant is nonvolatile the reaction rate increases both with the mass transfer and the wetted surface. If the limiting reactant is present also in the gas phase the gas covered surface is more efficient due to the higher mass transfer rate and the observed reaction rate decreases as the wetted fraction increases. Because of the competing effects between mass transfer resistance and the wetted fraction, the observed reaction rate may thus increase or decrease with the liquid flow rate as it is known that the mass transfer coefficient and the wetted external catalyst surface both increase with the liquid flow rate. Particulars of this general analysis are confirmed by Beaudry et al [326] and Tukač et al [327] who explain the decrease of the reaction rate by the inability of liquid rapidly to diffuse into the dry reaction zones.

Modification, or, better to say, expansion of the model of trickle bed reactor accounting for the flow inhomogeneities and the existence of the wall flow, presented in chapters 6.1 and 6.2, to incorporate rigorously all phase transitions shown in Fig. 6-14 is a formidable task. Nevertheless, this task is manageable with the aid of present powerful computers and appears worth undertaking in the light of its importance for trickle bed reactor design and elucidation of all features of its behaviour.

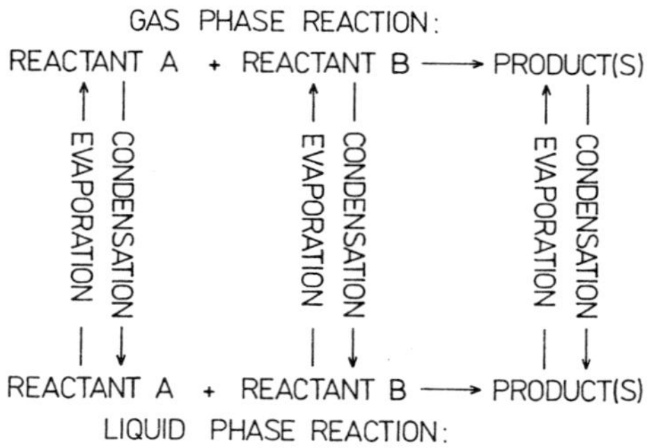

Fig.6-14
Scheme of the phenomena in a trickle bed reactor when the reaction takes place both in the liquid and the gas phase.

List of principal symbols to chapter 6

B, C	parameters of boundary conditions (6-1), (6-6), (6-11)
c, c_m, c_w	dimensionless concentrations of reacting species in liquid, cup-mixing mean concentration defined in Eq. (6-33), in wall flow, all as fractions of c_0
c_0	dimensional reference and inlet concentration of reactant in liquid
c_p,c_{pg}	specific heat of liquid and gas
D	coefficient of radial spread of liquid
Da_n	Damkoehler number for n-th order reaction defined in Eq. (6-17)
d_p	particle diameter
E	activation energy
f, f_0	dimensionless density of irrigation expressed as a multiple of mean density of irrigation f_0
H	dimensionless quantity defined in Eq. (6-1)
H_R, H_w	dimensionless quantities defined in Eqs (6-38) and (6-44)
ΔH	reaction heat
h	depth of catalyst layer
h_w	heat transfer coefficient into coolant
k	pre-exponential factor of reaction rate defined in Eq. (6-13)

K_R	effective radial thermal conductivity of bed
P	ratio of fluxes of thermal capacities defined in Eq. (6-39)
P_p, P_w	dimensionless quantities defined in Eqs (6-28) and (6-30)
r, r_1, r_0	dimensionless coordinate of radius, dimensionless radius, both as a fraction of reactor radius r_0
R, R_0	reaction rate per unit of reactor volume and reference value at c_0 and T_0
R_g	versatile gas constant
T', T	dimensional and dimensionless temperature defined in Eq. (6-10)
T_m, T_c, T_w, T_{rmax}	dimensionless temperatures of cup-mixing mean defined in Eq. (6-52), coolant, wall flow, maximum reaction rate defined in Eq. (6-57); relationship to dimensional temperatures analogous to Eq. (6-10)
T_0	reference and inlet dimensional temperature
U	dimensionless adiabatic temperature rise defined in Eq. (6-25)
V	superficial velocity of gas
W	dimensionless wall flow as a fraction of $\pi r_0^2 f_0$
x	dimensionless axial coordinate as a fraction of h
X_H	mole fraction of hydrogen in gas phase
γ	dimensionless activation energy defined in Eq. (6-10)
ϱ, ϱ_g	density of liquid and gas

Additional symbols defined in text.

7. Conclusion

The aim of this book has been to provide a unified view of the relations between structural inhomogeneities of a bed of particles and subsequently induced mutual effects of inhomogeneities and peculiarities of the velocity and pressure field in the fluids passing through this bed, transport phenomena and the course of chemical reactions. Since one- and two-phase flow are found in a number of processes, plants and equipment of a variety of technologies, the results obtained are valid generally without limitation to a specific research or industrial activity.

A primary requirement for the examination of the mutual effects of the velocity field, transport phenomena and chemical reactions in beds of particles permeated by fluids is the structure of the bed itself. Relations have been shown between the changes of the structure of the bed, on the one hand, and the area of the void fraction, on the other hand.

The immediate result of structural inhomogeneities is the inhomogeneous velocity and pressure field in the fluid passing through the bed. The consequences for the generally spatially variable resistances in a structurally inhomogeneous bed have been analyzed theoretically by formulating models using differential vectorial Ergun equation. Although in this form the equation retains the original Ergun's empirical elements its vectorial differential form permits solution of more general problems in inhomogeneous structures without enhancing the empiricism of the approach.

A significant practical contribution of the analysis is the finding that the flow through an inhomogeneous structure is from energy point of view less demanding than the flow through the homogeneous structure of equal integral resistance. Purposeful structuralization of the layer can thus provide energy savings whose greater part, however, is dissipated at the "interfaces" between zones of particles of different size. Even in the presence of these additional losses structuralization of beds permits extensive manipulation with the velocity profile of the fluid from markedly axial upwards to markedly peripheral flows. Structuralization of the bed can also dramatically increase the intensity of radial transport, which is often a limiting factor.

Structuralization of the bed of particles in order to affect energy losses, velocity field and correction of undesirable operational regimes has so far been exploited almost exclusively in blast furnace metallurgy. The reasons for this situation are more historical than rational. Potentials for structuralization of the bed in reactor engineering remain virtually untapped.

It has also been shown that the coupling between the structure of the velocity field and heat dispersion in the bed is strong, even in structurally homogeneous beds. Since the heat transfer is always associated with temperature gradients, thermally induced flow inhomogeneities constitute an inseparable part of the processes in beds of particles.

The effects of flow inhomogeneities in systems with chemical reactions are extraordinarily strong. The strongest coupling with the course of a heterogeneously catalyzed reaction in a trickle bed reactor was detected for a reaction whose order was close to zero. The interactions with the relatively narrow reaction zones are particularly strong in this area and the behaviour of the reactor can be neither correctly predicted nor interpreted unless all aspects of the interactions are properly accounted for.

A relationship has been demostrated between the appearance of the hot spot in the trickle bed reactor with the evaporation of the reaction mixture, on the one hand, and the quantities characterizing the inhomogeneity of the flow, effective thermal conductivity of the catalytic bed and additional quantities, on the other hand.

REFERENCES

1. Kolář V., Staněk V.: Distribution of liquid over random packing. Coll. Czech. Chem. Commun. *30*, 1054 (1965).
2. Staněk V., Kolář V.: Distribution of liquid over a random packing. II. Derivation of relations for the distribution of density of wetting in a cylindrical column. Coll. Czech. Chem. Commun. *33*, 2636 (1968).
3. Staněk V., Kolář V.: Distribution of liquid over a random packing. V. Dependence of the coefficient of radial spreading of liquid on the size and form of the packing element and on the mean density of wetting. Coll. Czech. Chem. Commun. *33*, 263(1968).
4. Staněk V., Kolář V.: Distribution of liquid over a random packing. IV. Verification of the boundary condition of liquid transfer between a packed bed and the wall of a cylindrical column, and evaluation of its parameters. Coll. Czech. Chem. Commun. *33*, 1062 (1968).
5. Staněk V., Kolář V.: Distribution of liquid over a random packing. III. Distribution of liquid in a low bed of packing wetted by a thin stream of liquid; determination of the coefficient of radial spreading of liquid Coll. Czech. Chem. Commun. *33*, 1049 (1968).
6. Staněk V., Tichý J., Kolář V.: Distribution of liquid over a random packing. VI. Effect of distribution of liquid on experimentally determined hold-up of liquid in randomly packed bed. Coll. Czech. Chem. Commun. *33*, 3235 (1968).
7. Staněk V., Szekely J.: Heat and gas flow interactions in nonisothermal packed beds. Part II. Systems with countercurrent gas-solid flow. Trans. of the Metallurgical Society of AIME *245*, 1175 (1969).
8. Staněk V., Szekely J.: Heat and gas flow interactions in nonisothermal packed beds. Part I. Fixed bed systems. Trans. of the Metallurgical Society of AIME *245*, 1175 (1969).
9. Staněk V., Szekely J.: The effect of gas and heat flow interactions on the performance of adiabatic packed bed reactors. Chem. Eng. Sci. *25*, 1149 (1970).
10. Staněk V., Szekely J.: The effect of non-uniform porosity in causing flow maldistribution in isothermal packed beds. Can. J. Chem. Eng. *50*, 9 (1972).
11. Staněk V., Kolář V.: A model of the effect of the distribution of liquid hold-up in a packed bed and a new concept of static hold-up. Chem. Eng. J. (Lausanne) *5*, 51 (1973).
12. Staněk V., Kolář V.: Distribution of liquid over a random packing. VIII. Distribution of the density of wetting in a packing for an arbitrary type of initial condition. Coll. Czech. Chem. Commun. *38*, 2865 (1973).
13. Staněk V., Kolář V.: Distribution of liquid over a random packing. VII. The dependence of distribution parameters on the column and packing diameter. Coll. Czech. Chem. Commun. *38*, 1012 (1973).
14. Staněk V., Szekely J.: Flow maldistribution in two dimensional packed beds. Part II. The behaviour of nonisothermal systems. Can. J. Chem. Eng. *51*, 22 (1973).
15. Staněk V., Kolář V.: Distribution of liquid over a random packing. IX. The radial spread and the number of rivulets in a trickle bed. Coll. Czech. Chem. Commun. *39*, 2007 (1974).

16. Staněk V., Szekely J.: Three-dimensional flow of fluids through nonuniform packed beds. AICHE J. *20*, 974 (1974).
17. Staněk V.: The effect of the spread of fluid on the top of the bed on radial spread within the packing. Coll. Czech. Chem. Commun. *42*, 1129 (1977).
18. Staněk V., Mrázek J.: The relation between energy requirements for the flow, flow maldistribution and the structure of the stack region of a blast furnace. Archiv für das Eisenhüttenwesen *48*, 563 (1977).
19. Staněk V., Kolev N.: A study of the dependence of radial spread of liquid in random beds on local conditions of irrigation. Chem. Eng. Sci. *33*, 1049 (1978).
20. Staněk V., Eckert V.: A stratified model of a random bed of equal-diameter spheres confined by a plane. Coll. Czech. Chem. Commun. *44*, 829 (1979).
21. Staněk V., Eckert V.: Course of porosity in layered beds of spherical particles. Zetschrift für das Eisenhüttenwesen *50*, 19 (1979).
22. Staněk V., Eckert V.: A study of the area porosity profiles in a bed of equal-diameter spheres confined by a plane. Chem. Eng. Sci. *34*, 933 (1979).
23. Staněk V.: Početní stanovení preferenčního toku tekutiny v blízkosti stěn náplňového zařízení. Chem. Prům. *30*, 50 (1980).
24. Staněk V., Hanika J., Hlaváček V., Trnka O.: The effect of liquid flow distribution on the behaviour of a trickle bed reactor. Chem. Eng. Sci. *36*, 1045 (1981).
25. Staněk V., Vychodil P.: Analiz vozhmostnosti sovemstnogo eksperimentalnogo operedeleniya koeffitsientov radialnoi i osevoi teploprovodnosti v nasadochnykh kolonakh. "Sovremennyie eksperimentalnyie metody issledovania protsessov teplo- i massoobmena", A. Soloukhin, V. A. Borodulyia, 1-st part, page *168* (1981).
26. Staněk V., Hanika J.: The effect of flow distribution on catalytic hydrogenation of cyclohexene in an adiabatic trickle-bed reactor. Chem. Eng. Sci. *37*, 1283 (1982).
27. Staněk V., Vychodil P.: On the observability of heat dispersion parameters in packed beds. Coll. Czech. Chem. Commun. *48*, 2484 (1983).
28. Staněk V., Kalousek V.: Course of porosity in layered beds of coke and agglomerate. Stahl und Eisen *54*, 133 (1983).
29. Hanika J., Lukianov B.N., Kirillov V.A., Staněk V.: Hydrogenation of cyclooctadiene in a trickle bed reactor accompanied by phase transition. Chem. Eng. Commun. *40*, 183 (1986).
30. Staněk V., Vychodil P.: On the length effect of the heat dispersion parameters. Chem. Eng. Commun. *27*, 69 (1984).
31. Staněk V., Vychodil P.: Fateiev G., A., Tarasevich V. L.: Calculation of the dynamics of thermal desorption of hydrogen from a layer of intermetallide. Coll. Czech. Chem. Commun. *49*, 1881 (1984).
32. Staněk V.: Matematický model procesu spékání částic ve vrstvě sypkého materiálu. Research report, 36 (1973) Inst. Chem. Proc. Fund., Czech. Acad. Sci., Prague 1973.
33. Staněk V.: Vzestupný tok plynu vrstvou sypkého materiálu. Research report 32 (1973), Inst. Chem. Proc. Fund., Czech. Acad. Sci., Prague 1973.
34. Staněk V.: Vzestupný tok plynu vrstvou sypkého materiálu. Research report 40 (1974), Inst. Chem. Proc. Fund., Czech. Acad. Sci., Prague 1974.
35. Staněk V.: Vzestupný tok plynu vrstvou sypkého materiálu. Research report 37 (1975), Inst. Chem. Proc. Fund., Czech. Acad. Sci., Prague 1975.
36. Staněk V.: Vzestupný tok plynu vrstvou sypkého materiálu. Research report 41 (1976), Inst. Chem. Proc. Fund., Czech. Acad. Sci., Prague 1976.
37. Staněk V.: Hodnocení zavážky vysoké pece z hlediska intenzifikace procesu. Research report 39 (1977), Inst. Chem. Proc. Fund., Czech. Acad. Sci., Prague 1977.
38. Staněk V.: Experimentální studie tlakové ztráty v šachtě vysoké pece. Research report 23 (1978), Inst. Chem. Proc. Fund., Czech. Acad. Sci., Prague 1978.

39. Staněk V.: Studie vztahu mezi strukturou vsázky a tlakovou ztrátou v šachtě vysoké pece. Research report 27 (1979), Inst. Chem. Proc. Fund., Czech. Acad. Sci., Prague 1979.
40. Staněk V.: Simulace vztahu struktury vsázky a režimu vzestupného toku plynu. Research report 46 (1980), Inst. Chem. Proc. Fund., Czech. Acad. Sci., Prague 1980.
41. Vychodil P., Staněk V.: Transport tepla tuhá částice plyn v pevných ložích. Research report 33 (1980), Inst. Chem. Proc. Fund., Czech. Acad. Sci., Prague 1980.
42. Staněk V.: Experimentální studie profilu porozity mezi vrstvami koksu a aglomerátu. Research report 48 (1981), Inst. Chem. Proc. Fund., Czech. Acad. Sci., Prague 1981.
43. Staněk V.: Početní stanovení vlivu dodatkových odporů mezi vrstvami koksu a aglomerátu. Research report 37 (1982), Inst. Chem. Proc. Fund., Czech. Acad. Sci., Prague 1982.
44. Wicksell S.D.: The corpuscle problem. A mathematical study of a biometric problem. Biometrika *7*, 84 (1925).
45. Debbas S., Rumpf H.: On the randomness of beds packed with spheres of irregular particles. Chem. Eng. Sci. *21*, 583 (1966).
46. Scott G. D.: Packing of spheres. Nature *188*, 908 (1960).
47. Bernall J. D., Mason J.: Coordination of randomly packed spheres. Nature *188*, 910 (1960).
48. Ridgway K., Tarbuck K. J.: Radial voidage variations in randomly packed beds of spheres of different sizes. Jour. of Pharmacy and Pharmacology *18*, 168S (1966).
49. Kondelík P., Horák J., Tesařová J.: Heat and mass transfer in heterogeneous catalysis. Variation of local void fraction in randomly packed beds of equilateral cylinders. Ind. Eng. Chem. Proc. Des. Develop. *7*, 250 (1968).
50. Ergun S.: Mass transfer rate in packed columns. Its analogy to pressure loss. Chem. Eng. Progr. *48*, 227 (1952).
51. Szekely J., Propster M.: Resistance of layer-charged blast furnace burdens to gas flow. Ironmaking and Steelmaking. No.1, pp.15 (1977).
52. Drahoš J., Čermák J., Ziolkowska O., Ziolkowski D.: Statistical analysis of local gas velocities at the exit from a packed bed. The Chem. Eng. J. *24*, 71 (1982).
53. Cihla Z., Schmidt O.: Studie o chování kapaliny při volném stékání výplní válcové věže I. Chemické listy *51*, 1 (1957).
54. Onda K., Takeuchi H., Maeda Y., Takeuchi N.: Liquid distribution in a packed column. Chem. Eng. Sci. *28*, 1677 (1973).
55. Dutkai E., Ruckenstein E.: Liquid ditribution in packed columns. Chem. Eng. Sci. *23*, 1365 (1968).
56. Zarzycki R.: Rozplyw cieczy w kolumnie wypelnionej. I. Mechanizm rozplywu cieczy w kolumnie wypelnionej. Inzynieria Chemiczna *1*, 85 (1971).
57. Porter K. E., Templeman J. J.: Liquid flow in packed columns. Part III: Wall flow. Trans. Inst. Chem. Engrs. *46*, T86 (1968).
58. Beveridge G. S. G., Haughey D. P.: Axial heat transfer in packed beds. Gas flow through beds between 20 and 650 °C. Int. J. Heat Mass Transfer *15*, 953 (1972).
59. Yagi S., Kunii D.: Studies on effective thermal conductivity in packed beds. AICHE J. *3*, 373 (1957).
60. De Wash A. P., Froment G. F.: Heat transfer in packed beds. Chem. Eng. Sci. *27*, 567 (1972).
61. Vortmeyer D., Schaefer R. J.: Equivalence of one- and two- phase models for heat transfer processes in packed beds: One dimensional theory. Chem. Eng. Sci. *29*, 485 (1974).
62. Vortmeyer D.: Axial heat dispersion in packed beds. Chem. Eng. Sci. *30*, 999 (1953).
63. Hodgson A. A. E.: Brit. patent 680 685 (1952).
64. Bonner L. D.: US patent 2 635 989 (1953).
65. Hanika J., Sporka K., Růžička J., Krausová J.: Qualitative observations of heat and mass transfer on the behaviour of a trickle bed reactor. Chem. Eng. Commun. *2*, 19 (1975).

66. Kitaev B. I., Jaroshenko Yu. G., Shrydkii V. S., Gordon Ya. M., Scherbatskii V. B., Shlyar B. A., Bokovikov B. A.: Investigation of the aerodynamics in packed metallurgical furnaces with tuyere wind entry. Blast Furnace Aerodynamics Symp. (ed. N. Standish), Australia, Australian Institute of Mining and Metallurgy. Ilawana Branch, Wolonglong, pp 68 (1975).
67. Yoshizawa A.: A proper tensorial form of Darcy-Ergun equation. Trans. ISIJ, *19*, 559 (1979).
68. Marquardt D. W.: An algorithm for least squares estimation of nonlinear parameters. J. Soc. Indust. Appl. Math. *11*, 431 (1963).
69. Hashimoto T., Muroyama K., Fujiyoshi K.: Radial effective thermal conductivity in gas-liquid cocurrent flow through packed beds. Kagaku Kogaku Ronbushu *2*, 53 (1976).
70. Muroyama K.: Heat transfer characteristics in gas-liquid cocurrent packed beds. Okayama, Japan, October 1978.
71. Crine M.: Heat transfer phenomena in tricklebed reactors. Chem. Eng. Commun. *19*, 99 (1982).
72. Staněk V.: Three-dimensional flow of fluids through nonuniform packed beds. Congress CHISA, Prague 1975.
73. Radestock J.: Untersuchung der stationaren inkompressiblen und kompressiblen Strömungen in ruhenden, geschichteten und isotropen Schütungen. Dissertation. Technische Universität Clausthal-Zellerfeld, 1969.
74. Radestock J., Jeschar R.: Über die Strömung durch die Hochofenschüttung. Stahl und Eisen *90*, 1249 (1970).
75. Radestock J., Jeschar R.: Theoretische Untersuchung der inkompressiblen und kompressiblen Strömung durch Reaktor-Schüttungen. Chem. Ing. Tech. *43*, 355 (1971).
76. Radestock J., Jeschar R.: Theoretische Untersuchung der gegenseitigen Beeinflussung von Temperatur- und Strömungsfeldern in Schüttungen. Chem. Ing. Tech. *43*, 1304 (1971).
77. Kalousek V.: Experimental measurement of the porosity profiles in a random bed of particles. M. S. Thesis. Institute of Chemical Technology, Prague 1981.
78. Zagrafskaia R. V., Karnauchov A. P., Fenelov V. B.: Globuliarnaia model poristykh tel korpuskularnogo stroenia. III. Issledovanie sluchainykh i chastichno uporiadochenykh sharov. Kinetika i kataliz *16*, 1583 (1975).
79. Choudhary M., Propster M., Szekely J.: On the importance of the inertial term in the modelling of flow maldistribution in packed beds. AICHE J. *22*, 600 (1976).
80. Fukutake T., Okabe K.: Theoretical and experimental studies of the scale up conditions for the gas flow in packed beds. Transactions ISIJ *16*, 189 (1976).
81. Brinkman H. C.: Calculation of the viscous force exerted by a flow of fluid on a dense swarm of particles. Appl. Sci. Res. *Section A1*, 27 (1947).
82. Lebon G., Cloot A.: A thermodynamical modelling of fluid flows through porous media: Application to natural convection. Int. J. Heat Mass Transfer *29*, 381 (1986).
83. Chandrasekhara B. C., Vortmeyer D.: Flow model for velocity distribution in fixed porous beds under isothermal conditions. Wärme- und Stoffübertragung *12*, 105 (1979).
84. Jeschar R.: Druckverlust in Mehrkornschüttungen aus Kugeln. Arch. für das Eisenhüttenwesen *35*, 1 (1964).
85. Abe E., Hirosue H., Yokota A.: Pressure drop through a packed bed of binary mixture. J. Chem. Eng. Japan *12*, 302 (1979)
86. Schwartz C. E., Smith J. M.: Flow distribution in packed beds. Ing. Eng. Chem. *45*, 1209 (1953).
87. Hennecke F. W., Schlünder E. U.: Wärmeübertragung in behitzten oder gekühlten Rohren mit Schüttungen aus Kugeln, Zylindern und Raschig-Ringen. Chem. Ing. Tech. *45*, 277 (1973).
88. Martin H.: Low Peclet number particle-to-fluid heat and mass transfer in packed beds. Chem. Eng. Sci. *33*, 913 (1978).
89. Fahien R. W., Stankovic J. M.: An equation for the velocity profile in packed columns. Chem. Eng. Sci. *34*, 1350 (1979).

90. Stroh K. R., Jiacoletti R. J., Olson H. G.: Thermal-hydraulic analysis techniques for axisymmetric pebble bed reactor cores. Nuclear Engng. Design *52*, 349 (1979).
91. Cohen Y., Metzner A. B.: Wall effects in laminar flow through packed beds. AICHE J. *27*, 705 (1981).
92. Vortmeyer D., Schuster J.: Evaluation of steady flow profiles in rectangular and cicular packed beds by variational method. Chem. Eng. Sci. *38*, 1691 (1983).
93. Vafai K., Alkire R. L., Tien C. L.: An experimental investigation of heat transfer in variable porosity media ASME Trans., J. Heat Transfer *107*, 642 (1985).
94. White S. M., Tien C. L.: Analysis of flow near the wall in packed beds. Wärme- und Stoffübertragung *21*, 291 (1987).
95. Szekely J., Poveromo J.: Flow distribution in packed beds: A comparison of measurement with predictions. AICHE J. *21*, 769 (1975).
96. Ziolkowska I.: Matematyczny model przeplywu gazu przez kolumne z bezladnie usypanym zlozem regularnych elementow. Inzynieria Chemiczna i Procesowa *1*, 641 (1980).
97. Ziolkowska I., Badowska I., Mieszkowski Z.: Matematyczny model przeplywu gazu przez kolumne z bezladnie usypanym zlozem regularnych elementov. II. Weryfikacja modelu na przekladzie zloz pierscieniu Raschiga i cylindrow. Inzynieria Chemiczna i Procesowa *1*, 889 (1980).
98. Ziolkowska I., Badowska I., Flejter B., Mieszkowski Z.: Wplyw wysokosti warstwy zloza na profil predkosci w rurze z wypelnieniem ziarnistym. Inzynieria Chemiczna i Procesowa *1*, 393 (1980).
99. Ziolkowska I., Badowska I., Morawska B.: Metoda wyznaczania skladowych wektora predkosci w kolumnie z wypelnieniem ziarnistym. Inzynieria Chemiczna *VI*, No.3, 693 (1976).
100. Kirrilov V. A., Kuzmin V. A., Khanaev V. M.: O profile skorosti v nepodvizhnom zernistom sloe. Doklady Akademii Nauk SSSR. *245*, 159 (1979).
101. Schuster J., Vortmeyer D.: Geschwindigkeitverteilung in gasdurchströmten, isothermen Kugelschüttungen. Chem. Ing. Tech. *53*, 806 (1981).
102. Price J.: The distribution of fluid velocities in randomly packed beds of spheres. Mech. Chem. Trans. *May 1968*, (1968).
103. Cairns J., Prausnitz J. M.: Velocity profiles in packed and fluidized beds. Ind. Eng. Chem. *51*, 1441 (1959).
104. Caycik M., Gunn D. J.: True mean velocities in the wall and bulk regions of fixed beds. Chem. Eng. Sci. *43*, 1637 (1988).
105. Van der Merwe D. F., Gauvin W. H.: Velocity and turbulence measurements of air flow through a packed bed. AICHE J. *127*, 519 (1971).
106. Lyczkowski R. W.: Modelling of flow nonuniformities in fissured porous media. Can. J. Chem. Eng. *60*, 61 (1982).
107. Ergun S.: Pressure drop in blast furnace and in cupola. Ind. Eng. Chem. *45*, 477 (1953).
108. Mickley H. S., Smith K. A., Korchak E. I.: Fluid flow in packed beds. Chem. Eng. Sci. *20*, 237 (1965).
109. Delmas H., Froment G. F.: A simulation model accounting for structural radial nonuniformities in fixed bed reactors. Chem. Eng. Sci. *43*, 2281 (1988).
110. Mac Donald I. F., El Sayed M. S., Mow K., Dullien F. A. L.: Flow through porous media. The Ergun equation revisited. Ind. Eng. Chem. Fundam. *18*, 199 (1979).
111. Foscolo P. U., Gibilaro L. G., Waldram S. P.: A unified model for particulate expansion of fluidized beds and flow in fixed porous media. Chem. Eng. Sci. *38*, 1251 (1983).
112. Kajiwara Y., Jimbo T., Sakai T.: Development of simulation model for burden distribution. Trans ISIJ *23*, 1045 (1983).
113. Narita K., Inaba S., Shimizu M., Yamaguchi A., Kobayashi I., Okimoto K.: Burden and gas distribution considering blast furnace aerodynamics. Trans. ISIJ *21*, 405 (1981).

114. Cheng P., Hsu C. T.: Fully developed, forced convective flow through an annular packed-sphere bed with wall effects. Int. J. Heat Mass Transfer *29*, 1843 (1986).

115. Pillai K. K.: Voidage vaiation at the wall of a packed bed of spheres. Chem. Eng. Sci. *32*, 59 (1977).

116. Govindaro V. M. H., Froment G. F.: Voidage profiles in packed bed of spheres. Chem. Eng. Sci. *41*, 533 (1986).

117. Benenati R. F., Brosilow C. B.: Void fraction distribution in beds of spheres AICHE J. *8*, 359 (1962).

118. Schuster J., Vortmeyer D.: Ein einfaches Verfahren zur naherungsweisen Bestimmung der Porosität in Schüttungen als Funktion des Wandabstandes. Chem. Ing. Tech. *52*, 848 (1980).

119. Thadani M. C., Peebles F. N.: Variation of local void fraction in randomly packed bed of equal spheres. I. E. C. Proc. Des. Develop. *5*, 265 (1966).

120. Kalny Z.: Metoda wyznaczania porowatosti lokalnej w kolumnie z wypelnieniem ziarnistim. Polish Scientific Conference on Chemical and Process Engineering, Warszawa, September 1977. Proceedings Vol. III. pp. 215-225.

121. Propster M., Szekely J.: The porosity of systems consisting of layers of different particles. Powder Technol. *17*, 123 (1977).

122. Schultz H. J., Abel O.: Schtttverhalten von Formkoks-Erz- und Stückkoks-Erz-Systemen. Stahl und Eisen *94*, 287 (1974).

123. Roblee L. H. S., Baird R. M., Tierney J. W.: Radial porosity variations in packed beds. AICHE J. *4*, 460 (1958).

124. Standish N., Borger D. E.: The porosity of particulate mixtures. Powder Technol. *22*, 121 (1979).

125. Gotoh K., Jodrey W. S., Tory E. M.: Variation of the local packing density near the wall of a randomly packed bed of equal spheres. Powder Technol. *20*, 257 (1978).

126. Ridgway K., Tarbuck K. J.: Voidage fluctuations in randomly packed beds of spheres adjacent to a containing wall. Chem. Eng. Sci. *23*, 1147 (1968).

127. Berninger R., Vortmeyer D.: Die Umströmung von Hindernissen in Schüttungen. Chem. Ing. Tech. *59*, 224 (1987).

128. Berninger R., Vortmeyer D.: Der Einfluß von Versperrung auf das reaktionstechnische Verhalten eines adiabaten Festbettreaktors. Chem. Ing. Tech. *60*, 1052 (1988).

129. Cumberland D. J., Crawford R. J.: The packing of particles. Handbook of Powder Technology, Vol. 6, Elsevier 1987.

130. Kubie J.: Influence of containing walls on the distribution of voidage in packed beds of uniform spheres. Chem. Eng. Sci. *43*, 1403 (1988).

131. Ouchiyama N., Tanaka T.: Porosity of a mass of solid particles having a range of sizes. IEC Fund. *20*, 66 (1981).

132. Dodds J. A.: The porosity and contact points in multicomponent random sphere packings calculated by a simple statistical geometrical model. J. Colloid and Interface Sci. *77*, 317 (1980).

133. Marivoet J., Teodoroiu P., Wajc S. J.: Porosity, velocity and temperature profiles in cylindrical packed beds. Chem. Eng. Sci. *29*, 1836 (1974).

134. Haughey D. P., Beveridge G. S. G.: Structural properties of packed beds - a review. Can J. Chem. Eng. *47*, 130 (1969).

135. Tory E. M., Church B. H., Tam M. K., Ratner M.: Simulated random packing of equal spheres. Can. J. Chem. Eng. *51*, 484 (1973).

136. Gotoh K., Jodrey W. S., Tory E. M.: A random packing structure of equal spheres - statistical geometrical analysis of tetrahedral configurations. Powder Technol. *20*, 233 (1978).

137. Kamel M. T., Tory E. M., Jodrey W. S.: The distribution of k-th nearest neighbours and its application to cluster settling in dispersion of equal spheres. Powder Technol. *24*, 19 (1979).

138. Jodrey W. S., Tory E. M.,: Computer simulation of isotropic, homogeneous, dense random packing of equal spheres. Powder Technol. *24*, 111 (1981).
139. Gotoh K., Finney J. L.: Statistical geometrical approach to random packing density of equal spheres. Nature *252*, 202 (1974).
140. Jodrey W. S., Tory E. M., Random Sequential packing in R. J. Statist. Comput. Simul. *10*, 87 (1980).
141. Haughey D. P., Beveridge G. S. G.: Local voidage variation in a randomly packed bed of equal-sized spheres. Chem. Eng. Sci. *21*, 905 (1966).
142. Haughey D. P., Beveridge G. S. G.: Local property variations in a randomly packed equal--sized spheres. Chem. Eng. Sci. *22*, 715 (1967).
143. Ziolkowska I., Ziolkowski D.: Fluid flow inside packed beds. Chem. Eng. Process. *23*, 137 (1988).
144. Larson R. G.: Derivation of generalized Darcy equations for creeping flow in porous media. Ing. Eng. Chem. Fund. *20*, 132 (1981).
145. Tondeur D., Kvaalen E.: Equipartition of entropy production. An optimality criterion for transfer and separation processes. Ind. Eng. Chem. Res. *26*, 50 (1987).
146. Sonntag G.: Einfluß des Lückenvolumens auf den Druckverlust in gasdurchstromten Füllkorpersäulen. Zeit. für technische Chem., Verfahrenstechnik und Apparatewesen *32*, 317 (1960).
147. Mitchell D. W.; Some aspects of airflow through sinter beds. J.Iron and Steel Inst. 198, 358 (1961).
148. Scheidig K., Gruson G.: Der Strömungwiderstand von Hochofeneinsatzstoffen. Neue Hütte *11*, 385 (1966).
149. Rumpf H., Gupte A. R.: Einflüsse der Porosität and Korngroßenverteilung im Widerstandgesetz der Porenströmung. Zeitschrift für Verfahrenstechnik, Technische Chemie und Apparatewesen *43*, 367 (1971).
150. Jeschar R., Potke W., Carlowitz O.: Durchgasungsprobleme bei der Direktreduktion in Schachtofen. Stahl und Eisen *99*, 904 (1979).
151. Jeschar R., Potke W., Petersen V., Polthier K.: Blast furnace aerodynamics. B. F. A. Symposium, Wollonglong, Australia, 1975, pp 136.
152. Brederhöft R., Jeschar R.: Druckverlust und optimale Koksstückgrobe beim Hochofenverfahren. Stahl und Eisen *93*, 55 (1973).
153. Schneider F. A., Rippin D. W. T.: Determination of the local voidage distribution in random packed beds of complex geometry. Ind. Eng. Chem. Res. *27*, 1936 (1988).
154. Dickinson E., Milne S. J., Patel M.: Structure and porosity of simulated sediments of polydisperse particles. Ind. Eng. Chem. Res. *27*, 1941 (1988).
155. Ouchiyama N., Tanaka T.: Porosity estimations of mixed assemblages of solid particles with different packing characteristics. J. Chem. Eng. Japan *21*, 157 (1988).
156. Bates M.P.: Aerodynamic model of the blast furnace. "Developments in the ironmaking practice" Meeting of the Iron and Steel Institute, London, November 1972, pp 677.
157. Schultz H. J., Abel O.: Durchströmungverhalten von Formkoks- Erz-Systemen. Arch. für das Eisenhüttenwesen. *45*, 279 (1974).
158. Schultz H. J., Abel O.: Druckverluste in Koks-Sinter-Schichtungen und Mischungen. Arch. für das Eisenhüttenwesen *45*, 445 (1974).
159. Kuwano Y., Suzuki K., Nakane C., Tate M.: On the limit of blowing rate of the blast furnace. Proceedings ICSTIS, Suppl. Trans. ISIJ *11*, 125 (1971).
160. Hoschide Y., Tagasaki S., Yunomura A.: Neue Entwicklungen in der Hochofentechnologie. Stahl und Eisen *100*, 1521 (1980).
161. Gudenau H. W., Kreibich K., Nomiya Y.: Einfluß unterschiedlich permeabler Zonen auf die Durchgasung im Hochofen unter besonderer Bercksichtigungder kohäsiven Zonen. Stahl und Eisen *100*, 1526 (1980).

162. Gudenau H. W., Kreibich K., Nomiya Y.: Modelluntersuchungen zur Hochofendurchgasung mit und ohne kohäsive Zone. Stahl und Eisen *100*, 488 (1980).
163. Gudenau H. W., Kreibich K., Peters K.H.: Messung und Optimierung des Profils kohäsiven Zone im Hochofen. Stahl und Eisen *101*, 147 (1981).
164. Szurman E.: Zusammenhang zwischen der Gasdurchlässigkeit des Hochofens und den mechanischen Kokseigenschaften. Stahl und Eisen *100*, 617 (1980).
165. Davidson J.: Burden testing and its application to blast-furnace operation. J. Iron and Steel Inst. *21*, 106 (1973).
166. Baker T., Chilton T. H., Vernon H. C.: The course of liquor flow in packed towers. Trans. Am. Inst. Chem. Engrs *31*, 296 (1935).
167. Tour R. S., Lerman F.: The unconfined distribution of liquid in tower packing. Trans. Am. Inst. Chem. Engrs. *35*, 719 (1939).
168. Tour R. S., Lerman F.: An improved device to demonstrate laws of frequency distribution. Trans. Am. Inst. Chem. Engrs. *35*, 709 (1939).
169. Tour R. S., Lerman F.: Area source distribution through unconfined packing. Trans. Am. Inst. Chem. Engrs. *40*, 79 (1940).
170. Kunesh J. G., Lahm L. L., Yanagi T.: Controlled maldistribution studies on random packing at commercial scale. I. Chem. Symp. Series No. 104, A233.
171. Perktold F.: Einfluß der Weberschen Schüttemethoden auf die Flüssigkeitausbreitung. Angew. Chem. *B19*, 184 (1947).
172. Chandrasekhar S.: Stochastic problems in astronomy. Rev. Mod. Phys. *15*, 1 (1943).
173. Porter K., Jones M. C.: A theoretical prediction of liquid distribution in a packed column with wall effect. Trans. Inst. Chem. Engrs *41*, 240 (1963).
174. Hoftyzer P. J.: Liquid distribution in a column with dumped packing. Trans. Inst. Chem. Engrs. *42*, T109 (1964).
175. Tepleman J. J., Porter K. E.: Experimental determination of wall flow in packed columns. Chem. Eng. Sci. *20*, 1139 (1965).
176. Huber M., Hiltbrunner R.: Füllkorperrektifizierkolonnen mit Maldistribution. Chem. Eng. Sci. *21*, 819 (1966).
177. Jameson G. J.: A model for liquid distribution in packed columns and trickle-bed reactors. Trans. Inst. Chem. Engrs. *44*, T198 (1966).
178. Jameson G. J.: Predictions from a model for liquid distribution in packed columns. Trans Inst. Chem. Engrs. *45*, T74 (1967).
179. Porter K. E.: Liquid flow in packed columns. Part I: The rivulet model. Trans. Inst. Chem. Engrs. *46*, T69 (1968).
180. Porter K. E., Barnett V. D., Templeman J. J.: Liquid flow in packed columns. Part II: The spread of liquid over random packings. Trans. Inst. Chem. Engrs. *46*, T74 (1968).
181. Hobler T.: Minimum zraszania powierzchni. Chem. Stos. *1B*, 145 (1964).
182. Hanika J., Prchlík J., Růžička V., Novák J.: Rozdělení dob zdržení při zkrápění neuspořádaného lože pórovitého katalyzátoru kapalinou. Chemický průmysl *23*, 551 (1973).
183. Soukup J., Kolomazník K., Zapletal V., Růžička V., Prchlík J.: Liquid distribution in trickle bed reactors in randomly packed porous beds. Coll. Czech. Chem. Commun. *38*, 3742 (1973).
184. Kolomazník K., Soukup J., Prchlík J., Zapletal V., Růžička V.: Experimental determination of the spreading coefficient Coll. Czech. Chem. Commun. *39*, 216 (1974).
185. Prchlík J., Soukup J., Zapletal V., Růžička V.: Liquid distribution in reactors with randomly packed porous beds. Coll. Czech. Chem. Commun. *40*, 845 (1975).
186. Prchlík J., Soukup J., Zapletal V., Růžička V., Kovařík P: Wall flow in trickle beds reactors. Coll. Czech. Chem. Commun. *40*, 845 (1975).
187. El-Din A. A. B., El-Halwagi M. M., Saleh M. A.: Liquid distribution in two-phase countercurrent packed column. Chem. Eng. Sci. *32*, 343 (1977).

188. Bemer G. G., Zuiderweg F. J.: Radial liquid spread and maldistribution in packed columns under different wetting conditions. Congress CHISA, Prague 1978.
189. Brignole E. A., Zacharonek G., Mangosio J.: Liquid distribution in packed beds. Chem. Eng. Sci. *28*, 1225 (1973).
190. Refre A. E., Hellincx: Flow conditions in a packed bed through local measurements. Chem. Eng. J. *18*, 1 (1979).
191. Crine M., Marchot P., L'Homme G. A.: Mathematical modelling of the liquid trickling flow through a packed bed using the percolation theory. Comp. Chem. Eng. *3*, 515 (1979).
192. Crine M., Marchot P., L'Homme G. A.: Liquid flow maldistribution in trickle-bed reactors. Chem. Eng. Commun. *7*, 377 (1980).
193. Farid M. M., Gunn D. J.: Liquid distribution and redistribution in packed columns - II. Chem. Eng. Sci. *33*, 1221 (1978).
194. Farid M. M., Gunn D. J.: Liquid distribution and redistribution in packed columns - I. Chem. Eng. Sci. *33*, 1211 (1978).
195. Jeschar R., Pötke W., Ehlert K. P., Mavroudis M.: Phänomenologische Modelversuche über die Flüssigkeitströmung im Unterofen des Hochofens. Stahl und Eisen *100*, 623 (1980).
196. Groenhof H. C.: Scaling-up of packed columns. Chem. Eng. J. *14*, 193 (1977).
197. Crine M., Marchot P.: Les ecoulements gaz-liquide a co-courant vers le bas en lit granulaire fixe. Premiere partie: Description des structures d'ecoulement du liquide. Entropie *17*, No 102 , 28 (1981).
198. Crine M., Marchot P.: Les ecoulements gaz-liquide a co-courant vers le bas en lit granulaire fixe. Seconde partie: Modele stochastique de l'ecoulement du liquide. Entropie *17*, No 102 , 37 (1981).
199. Szekely J., Kajiwara Y.: A mathematical representation of spatially non-uniform counter--current flow of gases and liquids in packed beds of relevance to flow phenomena in the bosh of iron blast furnaces. Trans. ISIJ *19*, 76 (1979).
200. Kolev N., Daraktschiev R., Markov L.: Labt sich die Randgangigkeit in Ftllkorperkolonnen beseitigen? Verfahrenstechnik *13*, 164 (1979).
201. Herskowitz M., Smith J. M.: Trickle-bed reactors: A review. AICHE J. *29*, 1 (1983).
202. Furzer I.A.: Liquid dispersion in packed columns - II. Chem. Eng. Sci. *39*, 1301 (1984).
203. Furzer I. A.: Liquid dispersion in packed columns - I. Chem. Eng. Sci. *39*, 1283 (1984).
204. Fox R. O., Fan L. T.: Statistical theory of liquid flow in trickle-bed reactors. World Congress III of Chemical Engineering, Tokyo, 1986, Reactor design 9b-157,pp 191.
205. Nakamura M., Mori H., Iwai K., Mizutani Y., Matsuzawa T., Toyama S.: Simultaneous measurement of radial distribution of gas and liquid flowrates in trickle bed reactors. World Congress III of Chemical Engineering, Tokyo, 1986, Reactor design 9b-305,pp 257.
206. Hoek P. J., Wesselingh J. A., Zuiderweg F. J.: Small scale and large scale liquid maldistribution in packed columns. Chem. Eng. Res. Des. *64*, 431 (1986).
207. Stikkelman R. M., Wesselingh J. A.: Liquid and gas flow patterns in packed columns. I. Chem. E. Symposium Series No.104, B155.
208. Kouri R. J., Sohlo J. J.: Liquid and gas flow patterns in random and structured packings. I. Chem. E. Symposium Series No.104, b193.
209. Stichlmair J., Stemmer A.: Influence of maldistribution on mass transfer in packed columns. I. Chem. E. Symposium Series No.104, B213.
210. Schlünder E. U.: Über den Einfluß von Phasenverteilungen auf den Wirkungsgrad von Trennstufen. Ber. Bunsenges. Phys. Chem. *83*, 1184 (1979).
211. Marchot P., Crine M., L'Homme G. A.: Two-phase flow through a packed bed. A stochastic model based on percolation concept. Chem. Eng. J. *36*, 141 (1987).
212. Ahtchl -All B., Pedersen H.: Very large lattice model of liquid mixing in trickle beds. Ind. Eng. Chem. *25*, 108 (1986).

213. Larson R. G., Scriven L. E., Davis H. T.: Percolation theory of two phase flow in porous media. Chem. Eng. Sci. *36*, 57 (1981).

214. Kunesh J. G., Lahm L., Yanagi T.: Commerical scale experiments that provide insight on packed tower distributors. Ind. Eng. Chem. Res. *26*, 1845 (1987).

215. Porter K. E., Jones M. C.: Liquid and gas distribution in the scale-up of packed columns. I.Chem. E. Symposium Series No.104, A245.

216. Zuiderweg F. J., Hoek P. J., Lahm L.: The effect of liquid distribution and redistribution on the separating efficiency of packed columns. I. Chem. E. Symposium Series No.104, A233.

217. Cihla Z., Schmidt O.: Studie o chování kapaliny při volném stékání výplní válcové věže. II. Chemické listy *51*, 1005 (1957).

218. Handbook of Heat and Mass Transfer. Vol.2: Mass transfer and reactor design. Edited by N. P. Cheremisinoff, pp 1049. Gulf Publ. Co., Houston 1986.

219. Bemer G. G., Zuiderweg F. J.: Radial spread and maldistribution in packed columns under different wetting conditions. Chem. Eng. Sci. *33*, 1637 (1978).

220. Albright M. A.: Packed tower distributor tested. Hydrocarbon processing *63*, No.9, 173 (1984).

221. Staněk V., Semkov K., Kolev N., Paskalev G.: Theoretical and experimental investigation of the wall flow deflecting ring. A single ring. Coll. Czech. Chem. Commun. *50*, 2685 (1985).

222. Semkov K., Kolev N., Staněk V., Moravec P.: Theoretical and experimental investigation of the function of the wall flow deflecting ring. A mathematical model for the case of a large number of deflecting rings. Coll. Czech. Chem. Commun. *52*, 1430 (1987).

223. Semkov K., Kolev N., Staněk V., Moravec P.: Theoretical and experimental investigation of the function of the wall flow deflecting ring. A generalized mathematical model for the case of a large number of deflecting rings. Coll. Czech. Chem. Commun. *52*, 1440 (1987).

224. Semkov K., Kolev N., Staněk V.: Theoretical and experimental investigation of the function of the wall flow deflecting ring. The determination of the optimum distance between deflecting rings. Coll. Czech. Chem. Commun. *52*, 2438 (1987).

225. Haidegger E., Vortmeyer D.: Simultane Lösung von Energie-, Stoff- und Impulsgleichungen für wandgekühlte chemische Festbettreaktoren. Chem. Ing. Tech. *61*, 647 (1989).

226. Molerus O., Schweinzer J.: Resistance of particle beds at Reynolds numbers up to Re = 10 . Chem. Eng. Sci. *44*, 1071 (1989).

227. Ziolkowski D., Szustek S.: Effect of fluid velocity radial profile on the mass dispersion in a fluid stream in a packed bed tubular apparatus. Chem. Eng. Sci. *44*, 1195 (1989).

228. Berl E.: U.S. Patent 1, 796, 501, March *17*, 1931.

229. Kirschbaum E.: Neue Fragen und Kentnisse aus Rektifikationstechnik. Z. Ver. deut. Ing. *75*, 1212 (1931).

230. Kirschbaum E.: German Patent 592, 906, February 1, 1934.

231. Lespinasse B., Le Goff P.: Recherches sur le absorption gaz-liquide. Rev. Inst. Fr. Pet. *17*, 21 (1962).

232. Charpentier J. C., Prost C., Swaaij W. P. M., Le Goff P.: Etude de la retention de liquide dans une colonne a garnissage arrose a co-courrant et a contre-courant de gaz-liquide. Genie Chim. *99*, 803 (1968).

233. Manning R. E., Cannon M. R.: Distillation improvement by control of phase channeling in packed columns. Ind. Eng. Chem. *49*, 347 (1957).

234. Mullin J. W.: The effect of maldistribution on the performance of a packed column. Ind. Chemist *33*, 408 (1957).

235. Gunn D. J.: Theory of liquid phase dispersion in packed columns. Chem. Eng. Sci. *35*, 2405 (1980).

236. Billet R.: Distillation engineering. Chemical publishing. New York, 1979.

237. Billet R., Mackowiak J.: EFCE Working party on Distillation, Absorption and Extraction Meeting. Helsinky, June, 1982.

238. Billet R., Schultes M.: Capacity studies of gas-liquid two phase countercurrent flow columns. I.Chem.Symposium Series No.104, B255.
239. Staněk V.: Unpublished results.
240. Kennedy C. R., Jaffe S. B.: Analysis of tracer experiments from commercial-scale trickle-bed reactors. Chem. Eng. Sci. *41* , 845 (1986).
241. Staněk V., Vychodil P.: Assessment of the role of thermally induced gas flow inhomogeneities in fixed bed reactors. Chem. Eng. Process. *24*, 203 (1988).
242. Krebs C.: Über den Einfluß der Diffusion und der Ungleichverteilung der Gasströmung auf die Trennwirkungs von Füllkorperkolonen. Chem. Ing. Tech. *55*, 303 (1983).
243. Krebs C.: Mass transfer in irrigated packed columns. Part II. The effect of diffusion and gas flow nonuniformities on the mass transfer. Chem. Eng. Process. *19*, 129 (1985).
244. Lin H., Horvát S.: Viscous dissipation in packed beds. Chem. Eng. Sci. *36*, 47 (1981).
245. Dixon A. G., DoConstanco M. A., Soucy B. A.: Fluid-phase radial transport in packed beds of low tube-to-particle diameter ratio. Int. J. Heat Mass Transfer *27*, 1701 (1984).
246. Melanson M. M., Dixon A. G.: Solid conduction in low d_t/d_p beds of spheres, pellets and rings. Int. J. Heat Mass Transfer *28*, 383 (1985).
247. Ziolkowski D., Legawiec B., Tobis J.: Wspolczynnik wnikania ciepla podczas ogrzewania strumenia gazu prez sciane rurowego aparatu z nieruchomym zlozem ziarnistym. Inzynieria Chemiczna i Procesowa *3*, 765 (1982).
248. Ziolkowski D., Tobis J.: Uproszony model celkowo dyfuzyjnyj przenosenia ciepla w cieczy przeplywajacej przez rorowy wimiennik ze zlozem ziarn kulistych. Inzynieria Chemiczna i Procesowa *4*, 249 (1983).
249. Moranville M. B., Kessler D. P., Greenkorn R. A.: Dispersion in layered porous media. AICHE J. *23*, 786 (1977).
250. Li C., Finlayson B. A: Heat transfer in packed beds - a reevaluation. Chem. Eng. Sci. *32*, 1055 (1977).
251. Specchia V., Sicardi S.: Modified correlation for the conductive contribution of thermal conductivity in packed bed reactors. Chem. Eng. Commun *6*, 131 (1980).
252. Botteril J. S. M., Denloye A. O. O.: A theoretical model of heat transfer to a packed bed or quiescent fluidized bed. Chem. Eng. Sci. *33*, 509 (1978).
253. Glicksman L. R., Joos F. M.: Heat and mass transfer in fixed beds at low Renolds numbers. ASME/AICHE 18-th National Heat Transfer Conference. 79-HT-92, San Diego, August 1979.
254. Carbonell R. G.: Flow non-uniformities in packed beds: Effect on dispersion. Chem. Eng. Sci. *35*, 1347 (1980).
255. Kunii D., Suzuki M.: Particle-to-fluid heat and mass transfer coefficients in packed beds of fine particles. Int. J. Heat Mass Transfer *10*, 845 (1967).
256. Lee C. J., Chou S. M., Shuh S. S.: Improved heat transfer in packed beds. Jour. of the Chinese Inst of Engineers *3*, 133 (1980).
257. Dixon A. G., Labua L. A.: Wall-to-fluid coefficients for fixed bed heat and mass transfer. Int. J. Heat Mass Transfer *28*, 879 (1985).
258. Chen B. H., McMillan A. F., Wang S. T.: An analytical solution for dispersion in packed beds with low bed-to-particle ratios. Chem. Eng. Sci. *38*, 257 (1983).
259. Beasley D. E., Clark J. A.: Transient response of a packed bed for thermal energy storage. Int. J. Heat Mass Transfer *27*, 1659 (1984).
260. Jeschar R., Bredehöft R., Mavroudis M., Neumann H.,Pätke W.: Wärme- und strömungstechnische Probleme des Hochofensverfahrens. Stahl und Eisen *91*, 665 (1971).
261. Vortmeyer D.: Die mathematische Modellierung von Reaktions- und Austauschprozessen in durchströmten Festbetten unter Berücksichtigung von ungleichmaßigen Strömungsverteilungen. Wärme und Stoffübertragung *21*, 247 (1987).

262. Yagi S., Kunii D.: Studies of heat transfer near the wall surface in packed beds. AICHE J. *6*, 97 (1960).
263. Morales M., Spinn C. W., Smith J. M.: Velocities and effective thermal conductivites in packed beds. Ind. Eng. Chem. *43*, 225 (1951).
264. Cheng P., Vortmeyer D.: Transverse thermal dispersion and wall channelling in a packed bed with forced convective flow. Chem. Eng. Sci. *43*, 2523 (1988).
265. Tstotsas E., Schlünder E. U.: Some remarks on channelling and on radial dispersion in packed beds. Chem. Eng. Sci. *43*, 1200 (1988).
266. Vortmeyer D.: Modelierung chemischer Festbettreaktoren. Dechema-Monographie Band 94, pp 79-96, Verlag Chemie 1983.
267. Lerou J. J., Froment G. F.: Velocity temperature and conversion profiles in fixed bed catalytic reactors. Chem. Eng. Sci. *32*, 853 (1977).
268. Tobis J., Vortmeyer D.: The near-wall channelling effect on isothermal constant pattern adsorption. Chem. Eng. Sci. *43*, 1363 (1988).
269. Tsotsas E., Schlünder E. U.: On axial dispersion in packed beds with fluid flow. Chem. Eng. Process. *24*, 15 (1988).
270. Schertz W. W., Bischoff K. B.: Thermal and material transport in nonisothermal packed beds. AICHE J. *15*, 597 (1969).
271. Kondelík P., Boyarinov A. I.: Heat and mass transfer in heterogeneous catalysis. XI. Wall effect in tubular catalytic reactor. I. Coll. Czech. Chem. Commun. *34*, 3852 (1969).
272. Lerou J. J., Froment G. F.: Estimation of heat transfer parameters in packed beds from radial temperature profiles. Chem. Eng. J. *15*, 233 (1978).
273. Gnielinski V.: Wärme und Stoffübertragung in Festbetten. Chem. Ing. Tech. *52*, 228 (1980).
274. Dietz P. W.: Effective thermal conductivity of packed beds. Ind. Eng. Chem. Fundam *18*, 283 (1979).
275. Cybulski A., Kawecki W.: Przenosienie ciepla w warstwach uformowanych z ziarn nieregularnych. Inzynieria Chemiczna *2*, 355 (1972).
276. Okazaki M., Yamasaki T., Gotoh S., Toei R.: Effective thermal conductivity for granular beds of various binary mixtures. J. Chem. Eng. Japan *14*, 183 (1981).
277. Paterson W. R., Carberry J. J.: Fixed bed reactor modelling. The heat transfer problem. Chem. Eng. Sci. *38*, 175 (1983).
278. Levec J., Carbonell R. G.: Longitudinal and lateral thermal dispersion in packed beds. Part I: Theory. AICHE J. *31*, 581 (1985).
279. Levec J., Carbonell R. G.: Longitudinal and lateral thermal dispersion in packed beds. Part II: Comparison between theory and experiment. AICHE J. *31*, 5981 (1985).
280. Vortmeyer D., Le Mong S.: Anwendung des Äquivalentzprinzipes zwischen Ein- und Zweiphasenmodellen auf die Lösung der Regeneratorgleichungen. Wärme und Stoffübertragung *9*, 29 (1976).
281. Kasparek G., Vortmeyer D.: Wärmestrahlung in Schüttungen aus Kugeln mit vernachlassigbarem Wärmeleitwiderstand. Wärme und Stoffübertragung *9*, 117 (1976).
282. Bauer R., Schlünder E. U.: Effective radial thermal conductivity of packings in gas flow. Part I. Convective transport coefficient. Int. Chem. Eng. *18*, 181 (1978).
283. Bauer R., Schlünder E. U.: Effective radial thermal conductivity of packings in gas flow. Part II. Thermal conductivity of the packing. Int. Chem. Eng. *18*, 189 (1978).
284. Sahimi M., Hughes B. D., Scriven L. E., Davis H. T.: Dispersion in porous media - I. One--phase flow. Chem. Eng. Sci. *41*, 2103 (1986).
285. Roizard C., Tondeur D.: Heat transfer in percolated fixed beds with temperature-dependent physical properties, compressive and dispersive effects. Int. J. Heat Mass Transfer *29*, 1021 (1986).

286. Dixon A. G., Creswell D. L.: Effective heat transfer parameters for transient packed-bed models. AICHE J. *32*, 809 (1986).

287. Matematicheskoe modelirovanie kataliticheskich reaktorov, Edited by Yu. S. Matros. Novosibirsk, "Nauka', Sibirskoe Otdelenie, 1989.

288. Wijngaarden R. J.: The scaling-up of cooled tubular reactors. Thesis. Twente University, 1989.

289. Schlünder E. U., Tsotsas E.: Wärmeübertragung in Festbetten, durchmischten Schüttgüttern und Wirbelschichten. Georg Thieme Verlag Stuttgart, New York 1988.

290. Ruß H. P.: Über den Wärmetransport in durchtrömten Schüttungen. Thesis. Institut für Thermische Verfahrenstechnik der Universitat Karlsruhe, 1988.

291. Tsotsas E., Martin H.: Thermal conductivity of packed beds: A review. Chem. Eng. Process. *22*, 19 (1987).

292. Dixon A. G.: Thermal resistance models of packed-bed effective heat transfer parameters. AICHE J. *31*, 826 (1985).

293. Tsotsas E., Martin H.: Warmeübertragung zwischen einer Wand und dispersen Gas/Feststoff--Systemen. Chem. Ing. Tech. *61*, 677 (1989).

294. Zumbrunnen D. A., Viskanta R., Incropera F. P.: Heat transfer through granular beds at high temperature. Warme und Stoffübertragung *18*, 221 (1984).

295. Wei J., Cwiklinski R., Tomuro J., Xiao J.: Temperature differences in a moving bed reactor. Chem. Eng. Sci. *42*, 1175 (1987).

296. Vortmeyer D., Berninger R.: Comparison of dispersion models for calculation of temperature profiles in packed beds. Ger. Chem. Eng. *6*, 9 (1983).

297. Vortmeyer D.: Packed bed models and convective flow. ASME; Proceeding of the 1988 National Heat Transfer Conference. HTD-Vol. 1, pp 755.

298. Schlünder E. U.: Heat transfer between packed, agitated and fluidized beds and submerged surfaces. Chem. Eng. Commun. *9*, 273 (1981).

299. Matros Yu. S., Chumakova N. A.: Mnozhestvennost stacionarnykh rezhimov v adiabaticheskom sloe katalizatora. Dokl. Akad. Nauk. SSSR *250*, 1421 (1980).

300. Boreskov G. K., Matros Yu. S., Klenov O. P.,Lugovskoi V. I., Lakhmostov V. S.: Lokalnye neodnorodnosti v sloe katalizatora. Dokl. Akad. Nauk SSSR *258*, 1418 (1981).

301. Schlünder E. U.: Heat transfer to packed and stirred beds from the surface of immersed bodies. Chem. Eng. Process. *18*, 31 (1984).

302. Bauer R., Schlünder E. U.: Die effektive Wärmeleitfähigkeit gasdurchströmten Schüttungen. VDI Forschungsheft 582, Düsseldorf, VDI Verlag 1977.

303. Martin H.: Wärmeübertragung an Feststoffpartikeln. Technische Mitteilung *77*, Heft 12, 599 (1984).

304. Balakrishnan A. R., Pei D. C. T.: Heat transfer in gas-solid packed bed systems. A critical review. Ind. Eng. Chem. Process. Des. Dev. *18*, 30 (1979).

305. Chiew Y. C., Glandt E. D.: Simultaneous conduction and radiation in porous and composite materials: Effective thermal conductivity. Ind. Eng. Chem. Fund. *22*, 276 (1983).

306. Martin H.: Neue Kentnisse über den Wärme- und Stoffübergang in durchströmten Haufwerken bei niedrigen Peclet Zahlen. Chem. Ing. tech. *50*, 116 (1978).

307. Vychodil P., Staněk V.: Proměření tepelné vodivosti vrstev částic molekulárního síta při různých průtocích plynu. Research report 22 (1986), Inst. Chem. Proc. Fund., Czech. Acad. Sci., Prague 1986.

308. Vychodil P., Staněk V.: Vývoj teploty ve vrstvě regenerovaného molekulárního síta. Research report 6 (1987), Inst. Chem. Proc. Fund., Czech. Acad. Sci., Prague 1987.

309. Fateev G. A., Tarasevich V.L., Staněk V., Vychodil P.: Raschot dinamiki protsessov perenosa pro khranenie vodoroda v svyazannom sostoianii. Inzhenerno-fyzicheskii zhurnal *57*, No 1, 100 (1989).

310. Chu C. F., Ng K. M.: Flow in packed tubes with small tube to particle diameter ratio. AICHE J. *35*, 148 (1989).
311. Itaya Y., Adachi K., Mizuna M., Hasatani M.: Heat transfer and heat storage characteristics of optically semitransparent-material packed-bed solar air heater. J. Chem. Eng. Japan *22*, 11 (1989).
312. Wijngaarden R. J., Westerterp K. R.: Do the effective heat conductivity and the heat transfer coeffcients at the wall inside a packed bed depend on chemical reaction? Weakness and applicability of current models. Chem. Eng. Sci. *44*, 1653 (1989).
313. Comiti J., Renaud M.: A new model for determining mean structure parameters of fixed beds from pressure drop measurements: Application to beds packed with parallelepidal particles Chem. Eng. Sci. *44*, 1539 (1989).
314. Goto S., Smith J. M.: Trickle-bed reactor performance. Part II. Reaction studies. AICHE J. *21*, 414 (1975).
315. Colombo A. J., Baldi G., Sicardi S.: Solid-liquid contacting effectiveness in trickle bed reactors. Chem. Eng. Sci. *31*, 1101 (1976).
316. Farid M. M., Gunn D. J.: Dispersion in trickle and two-phase flow in packed columns. Chem. Eng. Sci. *34*, 579 (1979).
317. Sicardi S., Baldi G., Specchia V., Mazzarino I., Gianetto. A: Packing wetting in trickle bed reactors. Influence of the gas flow rate. Chem. Eng. Sci. *36*, 226 (1981).
318. Mills P. L., Duduković M. P.: Evaluation of liquid-solid contacting efficiency in trickle bed reactors by tracer methods. AICHE J. *27*, 893 (1981).
319. Sicardi S., Baldi G., Gianetto A., Specchia V.: Catalyst areas wetted by flowing and semistagnant liquid in trickle-bed reactors. Chem. Eng. Sci. *35*, 67 (1980).
320. Sakornwimon W., Sylvester N. D.: Effectiveness factors for partially wetted catalyst in trickle bed reactors. Ind. Eng. Chem. Process Des. Dev. *21*, 16 (1982).
321. Chu C. F., Ng K. M.: Effective thermal conductivity in trickle bed reactors: Application of effective medium theory and random walk analysis. Chem. Eng. Commun. *37*, 127 (1985).
322. Baldi G., Gianetto A., Sicardi S., Specchia V., Mazzarino I.: Oxidation of ethyl alcohol in trickle bed reactors: Analysis of the conversion rate. Can. J. Chem. Eng. *63*, 62 (1985).
323. Beaudry E. G., Mills P. L., Duduković M. P.: Gas-liquid downflow, upflow and countercurrent flow in packed beds: Reactor performance for gas-limiting reaction. World Congress III of Chemical Engineering, Tokyo, 1986, Reactor design 9b-152, pp 173.
324. LaVopa V., Satterfield C. N.: Some effects of vapour liquid equilibria on performance of a trickle bed reactor. Chem. Eng. Sci. *43*, 2157 (1988).
325. Shapiro M., Brenner H.: Dispersion of chemically reactive solute in spatially periodic model of porous medium. Chem. Eng. Sci. *43*, 551 (1988).
326. Beaudry E. G., Duduković M. P., Mills P. L.: Trickle-bed reactors: Liquid diffusional effects in gas-limited reaction. AICHE J. *33*, 1435 (1987).
327. Tukač V., Mazzarino I., Baldi G., Gianetto A., Sicardi S., Specchia V.: Conversion rates in a laboratory trickle-bed reactor during oxidation of ethanol. Chem. Eng. Sci. *41*, 17 (1986).
328. Gutsche S., Wild G., Midoux N., Martin H.: Hydrodynamik und Wärmeübertragung in zweiphasig in Aufstrom durchströmten Festbettreaktoren. Chem. Ing. Tech. *61*, 733 (1989).
329. Satterfield C. N.: Trickle be reactors. AICHE J. *21*, 209 (1975).
330. Hofmann H.: Multiphase catalytic packed-bed reactors. Catal. Rev.Sci.Eng. 17, 71 (1978).
331. Crine M., L'Homme G.: Recent trends in reactor modelling of catalytic trickle-bed reactors, in the book: Mass Transfer with Chemical Reaction in Multiphase Systems. Sijthoff and Nordhoff Int.Pub., NATO Advanced Study Institute Series, 1982.
332. Lazzaroni C. L., Keselman H. R., Figoli N. S.: Trickle bed reactors. Multiplicity of hydrodynamic states. Relation between the pressure drop and liquid holdup. Ing. Eng. Chem. Res. *28*, 119 (1989).

333. Jiřičný V., Staněk V.: Transients of the hydrodynamics of counter-current packed bed columns. Chem. Eng. Sci. *45*, 449 (1990).
334. Hanika J., Staněk V.: Operation and design of trickle bed reactors in: Handbook of Heat and Mass Transfer Operations. Editor N.Cheremisinoff. Gulf Publ.Houston, 1985.
335. Zuiderweg F. J., Hoek P. J., Lahm L. L.: The effect of liquid distribution and liquid redistribution on the separation efficiency of packed columns. Distillation and Absorption'87 Brighton.
336. Olujic Ž., De Graauw J.: Appearance of maldistribution in distillation columns equipped with high performance packings. Chem. Biochem. Eng. Q. 3, 181 (1989).
337. Moore F., Rukovena F.: Liquid and gas distribution in commerical packed towers. Chemical Plants and Processing *1987*, No.8, 11 (1987).
338. Vortmeyer D., Dietrich K. J., Ring. K. O.: Comparison of two phase model predictions for adiabatic packed bed chemical reactors. Advances in Chemistry Series, Number 133, Chemical Reaction Engineering-II. Am. Chem. Soc. 1974.
339. Baron T.: Generalized graphical method for the design of fixed bed catalytic reactors. Chem. Eng. Prog. *48*, 118 (1952).
340. Yu W., Denn M. M., Wei J.: Radial effects in moving bed coal gasifiers. Chem. Eng. Sci. *38*, 1467 (1983).
341. Vortmeyer D., Winter R. P.: Die Bedeutung der Strömungverteilung für die Modellierung von chemischen Festbettreaktoren bei höheren Reynolds-Zahlen und ausgedehnten Reaktionszonen. Chem. Ing. Tech. *55*, 950 (1983).
342. Vortmeyer D., Winter R. P.: The impact of porosity and velocity distribution on the theoretical prediction of fixed bed chemical reactor performance. Comparison with experimental data. ACS Symposium Series 169, pp 49, Chemical Reaction Engineering 1982.
343. Choudhary M., Szekely J., Weller S. W.: The effect of flow maldistribution on conversion in catalytic packed bed reactor. Part I. Analysis. AICHE J. *22*, 1021 (1976).
344. Choudhary M., Szekely J., Weller S. W.: The effect of flow maldistribution on conversion in catalytic packed bed reactor. Part II. Experimental studies. AICHE J. *22*, 1027 (1976).

AUTHOR INDEX

SUBJECT INDEX